PORTUGAL
A BOOK OF FOLK-WAYS

BY THE SAME AUTHOR

A BOOK OF THE BASQUES

THE TRADITIONAL DANCE
(*with* VIOLET ALFORD)

OITO CANÇÔES REGIONAIS PORTUGUESAS

CANTARES DO POVO PORTUGUÊS

———

PORTUGAL
A BOOK OF FOLK-WAYS

by

RODNEY GALLOP

Illustrated
with photographs by the AUTHOR and
drawings by MARJORIE GALLOP

*Obra subsidiada pelo Instituto
de Alta Cultura*

CAMBRIDGE
AT THE UNIVERSITY PRESS

1961

PUBLISHED BY
THE SYNDICS OF THE CAMBRIDGE UNIVERSITY PRESS
Bentley House, 200 Euston Road, London, N.W. 1
American Branch: 32 East 57th Street, New York 22, N.Y.

First published 1936
Reprinted 1961

First printed in Great Britain at the University Press, Cambridge
Reprinted by offset-lithography by
Lowe & Brydone (Printers) Ltd, London, N.W.10

To the many Portuguese
in every province and in every walk of life
whose unfailing courtesy and kindliness
have helped me to write this book

Portingallers with us have troth in hand
Whose marchandie cometh much into England.
They are our friends with their commodities
And we English passen into their countries.

"The British Policy," Hakluyt, Vol. ii, p. 117

Contents

Illustrations   *page* ix

Preface   xi

Preface to the 1960 printing   xvi

PART ONE

WESTERN ARCADY

Chapter I THE SOUTH   *page* 3
       II THE NORTH   26

PART TWO

TRADITIONAL BELIEFS AND CUSTOMS

Chapter III MAGIC AND SUPERSTITION   *page* 49
        IV BIRTH, MARRIAGE AND DEATH   84
         V SPRING MAGIC   100
        VI SUMMER SAINTS   126
       VII THE DEATH OF THE YEAR   155

PART THREE

FOLK-MUSIC AND LITERATURE

Chapter VIII THE MUSIC OF FOLK-SONG   *page* 189
         IX THE TRADITIONAL BALLAD   213
          X THE POPULAR QUATRAIN   230
         XI THE FADO   245
        XII FOLK-TALES AND PROVERBS   266

BIBLIOGRAPHY   281

INDEX   285

Illustrations

PHOTOGRAPHS

Plate I A Fisherman at Nazaré   *facing page*
The *Capa de Honras* at Miranda do Douro   32

II Launching at Nazaré   33

III On the Douro   48

IV The Pig of Murça
Yoked Oxen at Oporto   49

V The Castle of Guimarães
São Mamede's Day at Janas   80

VI The Moorish Castle at Cintra   81

VII An Old Peasant   96

VIII *Entrada de Toiros* at Vila Franca da Xira
Carnival Costumes in Lisbon   97

IX *Mouriscos* at Sobrado
The Moors in *Floripes*   160

X The Dance of King David at Braga
The *Gota* at Carreço   161

XI Singing to the *Adufe*
Going to a *Romaria*   176

XII At a Prison Window
An Old Beggar   177

XIII Oil Jars at Estremoz   224

XIV A *Barco Saveiro* at Aveiro   225

XV A *Fado* in the street   240

XVI Coimbra   241

DRAWINGS IN THE TEXT

page		page	
xi	At Sezimbra	114	"Peixe corrido"
3	At a Ferra	117	In the Ribatejo
7	In Madragoa	121	Street Seller, Lisbon
10	"Alentejo não tem	126	Comes e Bebes
	Sombra"	129	N. Sra da Piedade da Mer-
15	In the Algarve		ceana
19	Campinos in the Ribatejo	134	Woman of Nazaré
23	Seaweed-gatherer at Apulia	139	Cargas at Mercês
26	A Vira on the Douro	141	Santo António on a Sardine
28	Tricana de Coimbra		Boat at Furadouro
34	Vintage	142	Alentejan Lavrador
36	Traditional Costume of the	150	Carding Pin
	Terras de Miranda	155	The Dança do Genebres at
39	Herdsman's Hut in Trás-		Lousa
	os-Montes	158	Arraial
41	The Vira do Minho	161	The Cirio da Quinta do
43	Traditional Costume of		Anjo
	Soajo	169	The Dança da Luta, Lisbon
49	Motif from a Carved Dis-	171	A Stick Dancer from Cercio
	taff (Trás-os-Montes)	173	The King of the Bugios
54	Weeding in the Alentejo	178	Princess Floripes
65	Motif from a Horn Cup	181	The Castle of Almourol
	(Alentejo)	189	Jogral and Dancer from the
71	The Brazier		*Cancioneiro da Ajuda*
75	Algarve Woman	193	Jogral from the *Cancioneiro*
78	An Alentejan Herd-boy		*da Ajuda*
79	Algarve Mule Cart	199	From the *Ordenações del*
84	At the Fountain		*Rei D. Manoel* (1514)
87	The Infant Jesus of Mir-	207	Douro Dancer
	anda do Douro	215	In Estremadura
89	Motif from an Embroi-	230	Market Day, Leiria
	dered Handkerchief	245	Fado de Lisboa
	(Viana do Castelo)	252	António Maneta, the Bag-
92	Barroso Woman		piper of Cabo da Roca
95	Alentejan Woman from	261	Fado de Coimbra
	Serpa	263	Repartee
100	The Estoril Turkey Man	266	The Stocking Cap
105	Carapaaaaaau Fresco!	269	An Old Varina
110	Carnival Mumming Play	271	Alentejan Blanket

Preface

Portugal is exceptionally rich in folk-lore. Its primitive population consisted of indigenous tribes who for lack of a better name may be called Lusitanians and who have been over-laid by later invaders and settlers; Greeks, Phoenicians, Celts, Romans, Goths, Moors and Normans, to say nothing of the many races which the busy intercourse of the Middle Ages brought to the Western shores of the Peninsula.

In the face of these successive incursions the country has remained predominantly pastoral and agricultural. Lisbon and Oporto are the only great cities. For the third place rival claims are put forward by Coimbra, Braga, Setúbal and Évora, all towns of less than thirty thousand inhabitants. A rustic countryside extends to the very gates of the two cities, and their immediate surroundings are more primitive, and the remoter regions less so than it might have seemed reasonable to expect.

Rich as the country is in folk-loric survivals, however, these are not to be met at every turn. Their pursuit may be compared with the less innocent occupation of birds-nesting. Those who are not actively on the lookout may roam the countryside day after day without noticing any nests. It is only those who deliberately go in search of them, parting the bushes and peering up at the sky through a network of branches, who find them. So it is with "popular antiquities". The traveller may spend weeks or months touring Portugal without stumbling upon many of these precious relics of the past. To discover them requires patience, diligent enquiry and above all some

notion of what to look for. They are to be observed only at
a particular combination of space and time. The *Mouriscada*
of Sobrado and the *Auto de Floripes* are not to be seen in every
village or on every day of the year. They survive as isolated
examples of something which may once have been general
but which is by no means so to-day.

It is one of the attractions of folk-lore that *plus ça change,
plus c'est la même chose.* However much it may reflect the local
colour and national character of each particular country, the
fundamental facts may be collated with those from other lands
to form a picture, which grows daily fuller and more convincing, of primitive modes of thought and ritual.

In Portugal a considerable accumulation of ethnographical
material has been garnered during the last fifty years and
printed in learned journals (some of them provincial) and a
few meritorious local monographs such as Padre Firmino
Martins' *Folklore de Vinhais* and Jaime Lopes Dias' *Ethno-
graphia da Beira Baixa.* Unfortunately little attempt has been
made to sift these discoveries and to review them in the light
of European research. Moreover, the work of collection has
been unevenly distributed. Traditional ballads and popular
quatrains have been transcribed in innumerable variants, but
little or no attention has been paid to the music to which
they are sung. Superstitious beliefs and practices have been
extensively recorded, while rural festivals and ceremonies
dependent on the calendar have been relatively neglected.

Neither in Portuguese nor in any other language has a
general review been attempted of the Portuguese folk-
heritage. Such therefore is the aim of the present volume.
Lack of time and opportunity have precluded a detailed study
of the decorative arts. These are briefly touched upon in the
first part of the book, which, written in the form of a rapid,
"personally conducted" tour of the country, is designed to
set the stage for the folk-activities with which the remainder
is concerned. This section may with equal advantage be read
first, last, or not at all. Nevertheless, since folk-ways, when
reduced to their barest essentials, have a marked similarity
throughout Europe, it seemed worth while to attempt a

description which might imbue them, in the reader's vision, with some of that vivid local colour in which the living reality is steeped.

The remainder of the book is a mosaic of materials compiled from countless different sources falling into three principal groups. In the first place I have drawn where necessary on materials already set down in print by reliable observers, foremost among them the veteran José Leite de Vasconcelos, doyen of Portuguese archaeology and ethnography, and his collaborators in the *Revista Lusitana*. Secondly, I am indebted to a large number of friends for unpublished information, communicated to me verbally and obtained, in many instances, at my own request. Thirdly, I have availed myself wherever possible of the fruits of first-hand observation and investigation. The proportion of this original contribution varies in the different chapters. Small in those dealing with folk-literature, where there is little scope left for original work, it grows progressively larger in those concerned with superstition, seasonal ceremonies and music.

In the second section, "Traditional Beliefs and Customs", I have chosen to postulate no general knowledge of folk-lore in my reader In order that it may be immediately intelligible to the uninitiated, I have gone over a great deal of ground already familiar to every folklorist, demonstrating anew, in the light of the Portuguese material, facts which he would be prepared to take for granted. In order, therefore, that the layman may fully apprehend the explanation of phenomena with which, in Portugal or elsewhere, he may already be familiar, the folklorist must seek compensation in the fresh material with which every step is illustrated.

In the nature of things folk-lore cannot be an exact science. Its conclusions can be demonstrated, but they cannot be proved. They repose on circumstantial evidence, on the balance of probabilities. My interpretations, therefore, are proposed, not imposed. To some they may prove acceptable. To others they may not. To these last I would plead that, if they have no other merit, they serve at least as a frame for this panorama of Portuguese folk-ways.

With regard to the third part, "Folk-Music and Literature", it will be clear that I have no great belief in the creative
originality of the folk. "All folk-cultures", I would say with
Padraic Colum, "are the popularisations of something that
was once aristocratic—music, poetry, costume, dance." In
the popular airs, verses, tales and proverbs of the Portuguese
I can see little more than the assimilation of themes from a
more cultivated milieu which have been transformed as they
have sunk through the varying strata of society. Tappert, in
his *Wandernde Melodien*, has put the matter in a nutshell.
"The folk", he writes, "cannot compose; it never creates...
it selects." Nevertheless, since it selects only what it finds
congenial, its choice expresses a racial spirit no less for its
lack of originality.

In Portugal, as everywhere else, the things of the folk are
rapidly disappearing. During the Middle Ages they suffered
intermittent ecclesiastical persecution. But no general decline
set in until the dawn of the ninet enth century ushered in the
era of liberalism. Must progress inevitably be accompanied
by some measure of vandalism? In Portugal old houses and
fortifications have been pulled down to make way for bandstands and municipal gardens. In the same way the old
occupations and recreations of the folk have been sacrificed
to a less regional civilisation. Mechanisation has speeded up
this process in the present century. Like pernicious weeds
the wireless, the sound-film and the gramophone are choking
the flowers of rural folk-song. Within the limits of my own
residence in Lisbon, the little altars of St Anthony which
survived among the children as a pretext for begging, were
sacrificed to the suppression of mendicity.

There is much, of course, of which one cannot regret the
passing. Dark superstitions, hanging like cobwebs in minds
obscured by ignorance and prejudice, lead to actions which
only too often have a tragic issue. It will doubtless be the
preoccupation of those in authority to sift the grain from the
chaff. While employing the twin instruments of punishment
and enlightenment to eradicate the more baneful practices,
they will seek with intelligent discernment to foster the

innocent pleasures of regional song and dance and rural festival. There is still time for these things to be saved from extinction or from the almost greater horrors of artificial revival.

Something has been done by private initiative. In ten or a dozen different localities groups have been formed among the peasants to preserve and perform their regional songs and dances. Activities such as those of Abel Viana at Carreço and Tomas Gomes Ciriaco at Serpa are worthy of every encouragement and support, if Portugal is not to follow other European countries into the dead level of drab mediocrity.

In conclusion, I must express my thanks to the many friends in Portugal who with assistance or hospitality have helped me to gather materials for this book, and in particular to the Condes de Aurora, Conde de Vilas Boas, Condesa da Castanheira, Condesa de Proença a Velha, Dona Mathilde Bensaude, Drs Raul Teixeira, Ferreira Deusdado, Busdorff Silva, Abel Viana, Srs Luis Chaves, Agostinho de Campos, Antonio Emilio de Campos, Jaime Lopes Dias, Tomas Gomes Ciriaco and João Bentes, the Misses Tait and Messrs William Tait, Aubrey Bell and Victor Delaforce, the Reynolds family and to the *Junta de Educação Nacional*.

Much of the material relating to dances has already appeared in *The Traditional Dance*, written in collaboration with Miss Violet Alford and published by Messrs Methuen to whom I am indebted for permission to reproduce it in a somewhat amplified form.

For the Portuguese passages quoted in the text I have endeavoured to adhere to the modernized system of orthography except in the case of proper names where there is an established English usage such as Cintra, Rocio, etc. I crave indulgence for any inconsistencies which may have escaped my eye.

R. G.

1936

Preface to the 1960 printing

This book stands unaltered as it was first written, a quarter of a century ago. The material for it was gathered in the years when, in Portugal as elsewhere, the old pattern of life in the countryside was beginning to break up under the impact of modern ideas and the folklorist had to work against time, as feverishly as the archaeologist snatching his sherds out of the path of the advancing bulldozer. When his career took him to Portugal, Rodney set himself to this task, equipped with a keen ear and eye, an enquiring mind, a lively interest in humanity—and a camera; but without the present-day folklorist's most useful tool, the tape-recorder. In the field of music, he had to rely on the difficult and exhausting business of noting down songs and dance tunes by ear. How easy, by comparison, is the collector's work today—dangerously easy, indeed, for the undiscriminating. But we can now enjoy the immense advantage of hearing, not only the songs, but the manner of singing them. Preservation for the specialist is one thing, presentation to a wider public without loss of their authentic quality is another, and Rodney, whose delight in the songs of Portugal made him want to share them with as many people as possible, was much troubled by this problem. He found an understanding interpreter in Eve Maxwell-Lyte, who sang them with the simplicity which he felt was essential, and together they did all they could to make Portuguese folksongs known in England.

"Folklore can never be an exact science" Rodney wrote in this book, but he always tried to bring scientific methods to bear on his research into custom and ritual. Like Miss Violet Alford, with whom he so often collaborated, he was very much aware of the limited outlook of those folklorists who (often through lack of opportunity) worked only in the narrow field of their own country or region, and claimed uniqueness for survivals which are found in varying forms all over Europe.

He constantly tried to break down these barriers, and to place the folklore of Portugal in a wider context.

The War put an end to this work, and his final illness prevented him from returning to it. Had he been able to do so, it is possible that he might have changed or modified some of his theories in the light of his own later experiences in other countries, or of more recent European research, for he was always as ready to accept fresh evidence from other people as to share his own knowledge with them. But since this was not to be, no attempt has been made to alter what he wrote in 1936.

The reprinting of this book was made possible by the support of the Instituto de Alta Cultura and the Secretariado Nacional de Informação. I would like to express my sincere thanks to them and to His Excellency Senhor Teotónio Pereira, who, while serving as Portuguese Ambassador in London, brought the idea to their attention, and whose interest in it was carried on by the Embassy staff after his departure. In Portugal, much valuable help was given by His Excellency Professor Dr Francisco Leite Pinto, Minister for Education, and Dr Diógo de Souza Holstein da Cámara Manoel. Above all, my warm and lasting gratitude is due to Rodney's friend and collaborator, Senhor António Emílio de Campos who first inspired the project, and continued to watch over it until its completion.

MARJORIE GALLOP

LONDON, 1960

Part One

WESTERN ARCADY

"...the Western Arcady that men call Portugal..."
GEORGE YOUNG

"Jardim da Europa a beira mar plantado"
(Garden of Europe planted by the sea)
RIBEIRO

CHAPTER I

THE SOUTH

Across an endless plain a train steams slowly in the crisp light of early morning. The plain is dun-coloured and desolate. Here and there, the even grain of its surface is smudged by thin, straggling woods of holm-oak, or graven in long, shallow furrows by a ploughman whose gaunt figure, carved out of the horizon, is the only sign of human life.

All is motionless and still, even the towering clouds which thrust their white mass up the sky, as clouds do in flat countries or near the sea. Away to the left, below the newly risen sun, a milky blue mountain range rims the plain, which elsewhere is bounded only by its own infinity. Along the hard line of the horizon, the slightest irregularity, intensified by the white light of Castile, assumes a disproportionate significance. A projecting rock, a track that oversteps the horizon, a scarcely perceptible curve drawn with the firm precision of a Persian miniaturist; these are the elements of a landscape limned with a truly classical reticence.

Now Ciudad Rodrigo glides past in the middle distance, cutting the blue sierra with the outline of its battlements from Romanesque cathedral to square-set castle. A mile farther on, the line crosses the River Agueda, its banks bordered with poplars which in autumn are scarlet against the blue of the slow-moving stream. Then the vast curtain of the sky drops on the bare stage where the traveller has been vouchsafed his last glimpse of dramatic, chevaleresque Spain. A little later, for a fleeting moment it lifts again, to reveal, ahead and to the left a glimpse of something new: three small, white triangles, like the topsails of a schooner, hull down upon the horizon. These are the snow-clad summits of the Serra da Estrela, the loftiest mountains in Portugal. A moment later they disappear below the grey-brown waves of the plain.

It is many hours before those mountains are again seen. The vast expanses of the plain crowd in once more upon the traveller, whispering to him of hallucination, denying the possibility of change. After an hour, two frontier stations are passed: Fuentes de Oñoro, of Peninsular War fame, then Vilar Formoso, the first village in Portugal. The grey-coated *carabineros* with shiny three-pointed hats give way to the blue-clad *guardas* of Portugal.

As the train proceeds, frontier formalities completed, the eyes of the traveller are fastened to the window avidly expectant of some change of scene such as may betoken a new country. For a moment, no more, he is disappointed. Then a miracle happens. The train quickens its pace. A fissure a few yards wide appears in the plateau, broadens and becomes a ravine. A stream appears from nowhere. All around, the plateau which seemed fixed in infinity and eternity is cracking, splintering, crumbling; and aided by its disintegration the train gathers speed. All of a sudden, the ground seems to open beneath the thundering wheels, and there appears, framed between the flanks of the ravine, a new world full fifteen hundred feet below. It is as though some daring mortal, venturing into a barrow of the Little People, had suddenly beheld the green land of Faery, glittering in sunshine, far below the surface of the earth.

Into this green distance the train slowly drops in wide, sweeping curves. The arid uplands give way to the pine woods, the rocky gorges, the heather-covered hills and flowering orchards of Beira Alta. The broad mass of the Estrela unfolds itself in all its majesty, the sun glittering on its high snow-fields. Busaco is passed, and also Coimbra and its white houses suspended over the river between the poplar groves of the Choupal. Every moment the landscape grows more smiling and more intimate, and the heart of the traveller rejoices in the blend of northern freshness and southern luxuriance which is the hall-mark of Portugal. Heather and bracken springing up among olive groves; hedges where the bramble is shouldered aside by knotted cactus and the periwinkle straggles at the foot of tall aloe spikes; purple pools of campanula and yellow splashes of lupin; liquid green of meadow grass and the metallic sheen of vines: these are the ingredients of a countryside which, in comparison with the austerity of Castile, inspires tender affection rather than admiration and respect.

Near the end of its course the line runs close to the Tagus. The white sails of invisible barges float like disembodied souls over the green marshes carrying with them memories of Holland or the Norfolk Broads, until with a rush and a roar and a choking eddy of smoke the train plunges through the highest hill of Lisbon and comes to rest in the Rocio station, where the platforms are on the second story.

It is a peculiarity of capital cities, almost, one might say, their royal prerogative, that they are pre-eminent by reason of their personality rather than of the public monuments and works of art which lie within their walls. What, for instance, would our English provincial cities be without their cathedrals, castles and colleges? Yet London, deprived of Westminster Abbey, St Paul's, of all its greatest edifices indeed, would still be London. So it is with Lisbon, to the charm and atmosphere of which its *Sehenswürdigkeiten* contribute comparatively little.

There are "sights", of course, which no traveller should miss. Black Horse Square, the Avenida, the spoilt romanesque cathedral, S. Vicente with the macabre attraction of its "pickled kings", the Museum of Coaches, the Rocio or "Rolling-Motion" square, shorn to-day of the waving mosaic pavement which won it this nickname from British sailors. And there are view points such as the Miradouro of Santa Luzia and the Senhora do Monte, overlooking serried ranks of brown roofs and tall, light-coloured façades.

But the charm of Lisbon is primarily human and only secondarily architectural. Never for one instant are its streets empty of people. Hence, with the help of the southern sunshine, the predominant impression which it leaves is one of seething vitality. Poverty here is accompanied by no air of bleak listlessness or growling discontent, but by an appearance of privations borne with gentle resignation.

In view of its position as an Atlantic port there is singularly little tang of the sea in Lisbon, save during the summer months when the cool *nortada* wind blusters through the city. Down by the waterside there are docks and quays, dark taverns and ship's chandlers' shops, but it is as though the sea and all that belongs to it recoiled impotently from the hills, one had almost said cliffs, on which the city is built. Despite the many cities "built on seven hills" few streets in Europe equal in steepness those of Lisbon. "Never did I behold", wrote Beckford, "such cursed ups-and-downs, such shelving descents and sudden rises, as occur at every step one takes in going about Lisbon."

The exploration of these hills is a fascinating pursuit. Each is a maze of narrow winding alleys, steep flights of steps and miniature squares, improbable intervals of horizontality, many of which are still known by ancient and picturesque names. There is the street of the Little Englishman, that of the Mother of Water, and one called with stark simplicity but with a hint of forgotten mystery, *Triste Feia*, the "Sad Ugly One". Nor did the young Republic, though it was no respector of names, feel constrained to rededicate to obscure demagogues the Alley of the Faithful of God and that of the Miracle of St Anthony.

The houses in these thickly populated quarters are solid and massive, though often sadly decayed. Obviously they have "come down in the world", and this note of vanished splendour is echoed in countless baroque portals and fountains where the women gather for gossip and to fill their glistening earthenware *bilhas*. Tall palaces, transformed into tenements, reveal their aristocratic features of whiteish limestone through peeling coats of *sang de bœuf* or old rose distemper, while others are diapered from head to foot with patterned tiles like some colossal German stove. The general impression of lightness is enhanced by the square dormer windows and the terra-cotta finials which give a coquettish upward curve to the steep roofs. There is an unreal, almost theatrical, quality in these Lisbon slums, especially those of Alfama and Mouraria which cling precariously to the slopes below St George's Castle. Here are streets, so narrow that they see none but the midday sun, yet gay with washing hung out like banners on a festal day.

One of the features of the city is its colony of fish-wives. *Peixeiras* and *varinas* (who take their name from the coastal town of Ovar whence they come) are always to be seen running  through the streets with flat baskets on their heads, balanced on little pads which in turn are supported by headgear resembling nothing so much as unextended opera hats. Their feet, traditionally bare, remain so, despite a municipal order to the contrary. They wear bright cottons, and 'kerchiefs knotted across their breasts, woven into intricate geometrical designs in black and red or green or the neutral tinge of French mustard. Their cries, some as musical as others are raucous, mingle with those of other itinerant vendors to form a never-ending symphony. Physically these women are magnificent. The *varinas* have Phoenician blood in their veins, and form, in the narrow alleys of Madragoa, a society apart which, thanks to their superior earning powers, is organised on a matriarchal basis.

The high parts of Lisbon look across the wide basin of the Tagus to a blue-green sea of pines, bounded by distant mountains against which the white church and houses of Almada stand out in sharp relief. To reach the Outra Banda (the name given to this part of the province of Estremadura) it is necessary to cross the Tagus to Cacilhas or Barreiro. So flat is the country lying to the east that in mid-stream the water stretches to a landless horizon both up and down the river.

From this point the finest prospect of Lisbon is to be enjoyed, that which Lady Mary Wortley Montagu found comparable only to Stamboul. It is lovely enough when the late light of afternoon throws clear-cut shadows across the jewelled facets of the houses. But I shall always recall it one pale, clear winter's dawn, when a low mist lay over the milky water and there floated above it, earthbound no longer, the three hills of St George, Bairro Alto and Estrela, opalescent and ethereal as a mirage. Through the mist the dark sails of anchored barges could be dimly discerned, hung out to dry in the windless calm. Lisbon that morning had the dreamlike quality of a Turner.

It is hard to imagine more delightful country than the Outra Banda. The Portuguese divide pine trees, like bulls, into "wild" and "tame". In the woods between Cacilhas and Azeitão the sombre green of the northern *pinheiro bravo* is an ideal foil to the glitter of the *manso* stone pine.

Set between the two castles, wrested from the Moors in the twelfth century, of Palmela and Sezimbra, the Serra da Arrábida is a little bit of Greece. There is, in particular, a hint of Souneion in the steep promontory of Espichel where the western escarpments of the range run into the sea. But the whole mountain is Greek in the ashen grey of its limestone ribs, in the blue translucent sea in which its summits are mirrored, and in the aromatic *maquis* with which the rock is overgrown. In May Arrábida blossoms with the passionate flowering of arid places. For a few weeks it is a rock garden of cistus, wild lavender, thyme, arbutus, juniper and a thousand other Mediterranean plants.

Then the scorching sun shrivels this transient beauty, and only a rough, hirsute growth clothes the steeper parts of the mountain. Lower down, where the Setúbal road winds among the foot-hills, the soil has lodged to form gentler slopes, chequered squares of green vineyard, cornland and red loam freshly turned by the plough. The first time, I wondered where I had already seen that contrast of purplish red and vivid green. Then I remembered the vineyards of Ionian Zante and the deep stains as of congealed blood where in August the currant grapes were set out to dry in the sun. The Greek impression was strengthened by a glimpse of Palmela, its white houses clustering high round a tawny stronghold like many an Aegean acropolis.

Setúbal is a largish town with fine old buildings, the amenities of which are not improved by its sardine-tinning industry. To reach it one passes through exquisite gardens

rich in fruit blossom and flowering shrubs, with the bright
green and gold of orange groves embroidered on a rich brocade
of cypress, olive and eucalyptus. Like a threatening sabre
drawn before the town, lies the long sandy promontory of
Troia in the sands of which lie buried the ruins of an ancient
city, Roman certainly and perhaps also Phoenician. On the
horizon the salt sea dunes pale away towards the south, and
the eye follows them into the incalculable distance where, at
long last, land and sea are one.

Above the water front and to the right, stands the fortress
of São Filipe, the sharp angle of its walls pointing out to sea
like the prow of a ship. Even thus must the Ark have looked,
stranded high and dry upon the summit of Ararat. At its foot
a road curls up past miniature coves of red sand across the face
of a sandy cliff where the dry bone of the earth shows through
the green flesh of gorse and broom. Back over Setúbal the
fortress of Palmela is outlined against the eastern sky, its three
brown towers echoed on lower levels by three white windmills and a lone pine tree, solitary upon the horizon. Below the
road, a square redoubt juts out into the waves, a purple judas
tree outlined against the soft grey masonry. In a walled-in
orange grove children are playing with a barking dog. A tall
eucalyptus stands sentinel over three boats drawn up on the
sandy beach.

Beyond Setúbal, except for a few low hills in the direction
of Montemor and Grândola, the country is of an unbroken
flatness. Although the bounds of Estremadura extend east-
wards to Vendas Novas and beyond Sines to the south, this
is to all intents and purposes the beginning of the Alentejo,
the largest province in Portugal, described by Caesar as the
Sicily of Hispania, and still, in the Middle Ages, the granary
of the Peninsula. So flat is this country that the very word
monte has lost the meaning of hill and has come to be applied
to the long, low whitewashed farmhouses with chimneys as
high again as themselves.

To journey here, where each farm, each ploughman or
shepherd that one passes becomes an incident of travel, has
a curiously quickening effect upon the perceptions. Alentejo

atones for any lack of variety by a spirit of place deeper and more satisfying than the superficial charm of hill and valley. When all the mountains of Europe have been debauched by tourist syndicates, these lowlands will still be able to call their soul their own. There is an austere majesty in the vast *charnecas* of cistus scrub where the lynx still runs wild; in the yellow cornlands and straggling woods of ilex and cork, the stripped trunks lean and brown as the bare shanks of the Alentejan children. On these desolate steppes, ancient, fortified cities profit by any slight eminence to lift castles, brown as the plain itself, over the white houses clustering at their feet like islands.

These cities glisten twenty miles away over the flat waves of the plain. One of them is Arraiolos, home of carpets woven in blue and a red which mellows with the years to softest bistre; home, too, of the famous *noiva de Arraiolos*, the legendary bride who took a fortnight to adorn herself and then appeared at the wedding wrapped only in a shepherd's cloak. Other cities are Montemor, Beja, Elvas of the plums and Estremoz. This last is famous for its potters, and it is well worth while to penetrate into one of the small dark sheds where the red-brown earthenware is made.

A foot pedal sets the potter's wheel swiftly turning. His hands constrict the shapeless mass of clay which rises in swift waves beneath his touch, swells in a bellying curve, draws in, and obedient to a final thumb-thrust curls over to form a flared-out lip. A moment later a deft cut with a piece of string severs the completed jar from the unformed clay and endows it with separate existence.

In many parts of Portugal the potter's craft is practised on a small scale, in the homes of the peasants. The Estremoz ware may be identified by the simple designs scratched on the smooth terra-cotta surface. Similar designs are engraved in Trás-os-Montes and Beira Alta on the so-called "black pottery" which, at its finest, resembles gun-metal or even pewter. At the village of Bisalhães, miniature pots of this ware are made for sale at a special fair held at Vila Real on St Peter's Day. Perhaps the most attractive pottery in Portugal is that made from Coimbra to Caldas da Rainha. On a cream or oatmeal-coloured glaze free floral patterns are stencilled (till lately they were drawn freehand) in blue, green, yellow, Venetian red, lavender and other hues. From Caldas are the huge dishes, some nearly two feet in diameter, enamelled with a high glaze of jade green, or, as once I saw, celadon. The red clay of the Mafra pottery is daubed with a slip of green, brown or buff, or combinations of these colours merging into one another, often applied in a drip pattern or decorated with small cream-coloured circles, and partially overlaid with a translucent glaze. Besides pots, jars and dishes, this ware includes pigs, bulls and fishes roughly fashioned in the round, and the figure of a woman whose wasp waist and bare breasts hint at Minoan influence, a feeling which is strengthened by the fact that she holds in her hand, in place of the serpent of the Minoan goddesses, a *sardão* lizard.

The chief glory of the Alentejo is the cathedral city of Évora, in the fifteenth and sixteenth centuries a great centre of art and music. The Portuguese are proudest of its Roman associations, for Évora is a namesake of our own York. The very beer is styled *Eborense*, and the local liqueurs are of the *Sertoriano* brand. But the Corinthian columns of the Temple of Diana served till 1834 as a slaughter-house, and the mighty aqueduct founded by Sertorius has been overgrown by the mediaeval city as a deserted castle by creeping ivy. On the outskirts, impudent little houses settle themselves between its straggling legs. As it thrusts its way up into the city they jump upon it

and drag it down to their own level, until at last it survives
only as a series of airy arches overstepping each successive
street and disappearing from one house into another.

  Although there are few concrete vestiges of Moorish art in
Évora there is something very mauresque in the general aspect
of the city; in the maze of narrow streets, the whitewash thick
as icing-sugar and the complexity of the buildings, haphazard
agglomerations of roofs, terraces, cupolas, buttresses and
balustrades of lace-like brickwork. Here, better than anywhere
else, one may trace the transition from the double horseshoe
of the Moors to the ornate Manoeline arch.

  All round Evora extends the tawny vastness of the plain.
Green in spring, yellow in summer, brown in autumn, it
follows with the changing season the hues of peasant pottery.
One has the feeling that only with an effort does the city keep
its head above the surface. In this country where one may walk
for a day through an unchanging landscape, where Nature
herself moves with ponderous deliberation, the peasantry have
a leisurely dignity of movement and demeanour. Nowhere in
Portugal are the people more firmly attached to the soil.
Emigration is almost unknown. There is line but little colour
in their costumes. Near Niza in the north, where the people
are of a patriarchal austerity, the sombre black garb of men
and women alike makes a wedding procession scarcely distinguishable from a funeral. Elsewhere the farmers retain the
short frogged jacket of green or brown cloth, the skin-tight
trousers of the *lavrador* (landowner), and wide-brimmed black
hats, some stiff like those of Andalusia, others soft and pliant
so that with time and use they take on different but uniformly
graceful shapes. Shepherds and drovers wear *ceifões* (chaps)
and *samarras* (open sleeveless jackets) of sheepskin, and wrap
themselves in striped rugs of undyed wool, brown and creamy
white, matching the flocks they guard. The women's dress is
less characteristic. Working in the fields in the broiling sun
they wear black felt hats like the men and, pulling their cotton
skirts up to the knees, fasten them between their legs like
bloomers. The *montes* that are their homes bear curious names
such as *Sempre Noiva* (Always a Bride), *Troca Leite* (Change

Milk), *Pouca Roupa* (Little Linen), and *Mata Mouros* (Slay
Moors) which is echoed in descending degrees of heroism by
Mata Lobos (Slay Wolves) and *Mata Burros* (Slay Donkeys).

Like all poor folk in Portugal, the Alentejan peasants live
with great simplicity. Only on feast days do they eat meat,
chicken or eggs. All that they can produce is sold in the market.
In the Serra de Monfurado a labourer's dinner or supper
consists of vegetable soup and a scrap of garlic sausage varied
with macaroni or beans, kept in home-made Thermos con-
tainers of cork. Throughout the Alentejo, work in the fields,
olive groves and cork woods is organised on traditional lines.
The cork stripping, in particular, is a highly skilled operation.
With easy virtuosity the *tiradores* hack through the bark with
their double axes without injuring the tender core of the tree
and peel it off as clean as a cast-off snakeskin. The *ganhão* or
locally hired labourer, wandering from *monte* to *monte*, is jack
of all trades except reaping. For the corn and olive harvest
bands (*maltas*) of labourers are recruited locally (*malteses*) or
come down from Beira (*ratinhos* or *bimbos*). Their foreman
(*maioral* or *menageiro*) is obeyed with the strictest discipline
in his administration of the rules, some of them very curious,
of their association. They have, for instance, a complicated
system of fagging and of privileges by seniority in the *malta*.
An elaborate ritual centres round the midday meal. This is
served in a huge cauldron. The men stand round with their
hats on and spoons in their hands. The *maioral* takes the first
spoonful, after which the place where he has dipped his spoon
is left to the last. The others advance in turn, help themselves,
take three steps backwards and eat. No complaints are allowed,
and every scrap must be finished.

To reach the Algarve one must cross a welter of low,
crumpled hills and drop down to the narrow strip of lowland
fringing the southern sea. The rulers of Portugal used to be
styled "King of Portugal and of the Algarves"; and there
were actually two Algarves, the one in Europe and the other
on the north coast of Africa. Only the first of these retained
the name, which is the Moorish Al-Gharbi, meaning the

"Kingdom of the West", though the Moors themselves called it Chenchir. It stretches from Vila Real de Santo António on the Guadiana to Cape St Vincent, the uttermost limit of Europe. Vila Real is a dull modern town from which the traveller looks across with envy at the marshmallow picturesqueness of Spanish Ayamonte. But it is the best centre from which to visit the Sotavento, the eastern half of Portugal's smallest province.

The Moors remained longer in the Algarve than in any other part of the country. They had been there for five hundred years before the treacherous surrender to Afonso III of the capital Xelb (Silves, which as a centre of Arabic culture ranked only second to Cordova) resulted in their final expulsion in 1242.

Nevertheless, the impression which the Moors left was not an abiding one, and the Algarve proves far less African in atmosphere than report and expectation would have it. Few of the little whitewashed towns lying along the lagoon-fringed coast have the flat-terraced roofs of Andalusia. Olhão, the most eastern in appearance, owes its terraces and lattice windows to no Arab heritage (for it was not founded till the sixteenth century) but to its modern commerce with North Africa. Faro, the capital, ravaged by the Earl of Essex in the sixteenth and by earthquake in the eighteenth century, is disappointing.

More attractive than the towns is the Algarve countryside. The main road runs parallel to the sea between gardens, vineyards and orchards of orange, lemon, olive and fig. *O Algarve é pai do figo* runs an old adage, but if it is the father of the fig it is the child of the almond, and in early February, when the weather is generally pleasant, it is worth coming many miles to see the endless sea of tenuous blossom which stretches behind Faro and Tavira.

Some miles back from the coast run the smoke-blue mountains of Malhão and São Miguel, starred with white farmhouses, long, low buildings, each crowned with a tall decorative chimney of tile and plaster. All the pride and artistic taste of the peasants is put into these chimneys which, like slender white campaniles, lend every house the air of a chapel. In the plain there is often a miniature aqueduct, linking the water wheel with the house and with the garden where the young tomatoes are protected from the frost by aloe leaves bent double and spiked into the ground.

There is no local costume, but the almond-eyed women, riding into town on mule-back, or filling their shapely water jars at the well, wear broad hats of heavy black felt over dark 'kerchiefs, a curiously becoming headgear.

At Loulé one enters the Barlavento, the western half of the province. The first place of interest is Silves, set on a fortified hill, where one may still see the cathedral which was once a mosque and the Gate of Treachery, which, according to the legend, was opened to Afonso's soldiers by a Moorish maiden.

Portimão, lying close by on the coast, is less interesting, but behind it rises the Serra de Monchique, deep blue from a distance, and green, as one approaches, with forests of cork and carob.

Near Portimão, too, is the Praia da Rocha, a golden beach girt with ruddy cliffs, marred unfortunately by a collection of hideous villas. As one follows the coast westwards past Lagos, a fresh, white little town, the soil grows more barren, and the landscape more arid. The distant *serras* fall away behind a foreground of low treeless hills. Soon, only the fig trees are left, and these bow ever lower before the strong winds from the west. At Vila do Bispo they grow to less than a man's height and trail their bare, bleached branches along the ground and even under it, looking, in their ordered rows, like a *corps de ballet* curtseying with outspread skirts.

At last, even the fig trees fail, and there is nothing but cistus scrub and stone. The contours flatten out, and the twin promontories of Sagres and St Vincent come into view, linked by a wide amphitheatre of vertical cliffs, stabbing the grey Atlantic. St Vincent, the extreme south-westerly corner of Europe, is marked only by a lighthouse. It is round the other headland that legend and tradition have gathered. In classical times Sagres was the Sacred Cape to which the gods came at night to rest from their journeyings. In Moorish times it was sanctified by a Christian church, and it was here that the relics of St Vincent came to shore guided and guarded by two holy crows.

But to the Portuguese the most sacred associations of Sagres are those which connect it with Prince Henry the "Navigator", son of the English Philippa of Lancaster. Here it was that this ascetic yet strangely fascinating dreamer passed his days in solitude, planning the voyages of discovery which made Portugal one of the pioneer colonial empires of Europe. It is easy to understand the appeal which the place exercised over him. Stark, barren, unbelievably desolate, Sagres appears the very brink of the known world, "where the last tangled edge slopes down to the abyss".

From the ruins of the Vila do Infante, and the chapel where the discoverers of Madeira dedicated the first roses brought from Africa, the traveller of to-day looks down upon the steamers which pass close in shore as though to bid a last farewell to Europe before setting their course for Africa. So also must this prince, who dedicated his whole life to the cause of discovery, have followed with his eyes the first voyagers who on his inspiration set forth into the unknown.

Barer than the Outra Banda, the part of Estremadura immediately north of Lisbon has a spacious beauty of its own. Behind the "royal village" of Belém with its memories of Vasco da Gama, and the modern hotels and villas of the Estorils lies a wide expanse of rolling moorland, starred in spring with dwarf gorse and yellow iris, from the midst of which rises the perennial miracle of Cintra. The Serra de Cintra is the upturned toe of a recumbent giant whose head and shoulders are the Serra da Estrela, while the mountains of northern Estremadura are his torso, his knees the Montejunto and his legs the lines of Torres Vedras.

By some freak of nature this rocky range contrives to draw all the moisture out of the surrounding plain, and there is always water dripping from the trees or running in little rills down the hillside. It has been described by a Portuguese poet as "the place where Winter has chosen to spend the Summer". There are days in July and August when the only cloud in an otherwise clear sky is the humid mantle which settles on Cintra, blotting out its jagged outline and nourishing the exuberant vegetation of its woods and gardens. Something of this almost tropical luxuriance is reflected in the architecture of the place, notably in the florid Manoeline of the Moorish Palace and the exotic pavilion of Beckford's Montserrat. The precipitous, pine-clad crags recall steel engravings of the Romantic period, and this romantic note, which seems to be inseparable from the residences of nineteenth-century royalty, is sounded with perhaps undue emphasis in the mock-mediaeval *Schloss* which crowns the Pena peak.

From the heights of Pena and Peninha the surrounding country looks flat, but this appearance is deceptive. It is a landscape of sweeping downland, refreshing and exhilarating in the clear, crisp light of winter, when the curving lines of newly turned furrows lend a solid, three-dimensional quality to the swelling contours of the hills. The plough is the very symbol of this landscape, and the endless monotonous chant of the ploughman to his beasts is its aptest *leitmotiv*.

At Mafra, behind the monstrous monastery with which João V sought to out-escorial Phillip II, the hills begin on which the famous lines of Torres Vedras were dug. Their clean, rhythmic outlines weave wonderfully satisfying patterns along the horizon, cut only by the rows of windmills, which crawl along the ridges, four or five at a time, like little white snails along the top of a wall.

Among these hills lie many country houses and *quintas*, or wall-enclosed properties, most of which date from the palmy days of the early eighteenth century. For all its Brazilian gold, for all its imitation of the sophisticated modalities of Versailles, the eighteenth century in Portugal was never completely divorced from the real countryside. Country life, in consequence, never quite lost the Theocritan charm of the *romaria* to degenerate into the stilted mannerisms of the *fête champêtre*. Thus the baroque style, in all its manifestations, was to some extent tempered by a saving simplicity. Two centuries later, the passage of time has, as it were, integrated with Nature herself an Arcadianism which was never entirely artificial. Eighteenth-century buildings, usually so incongruous in rural surroundings, have merged into the countryside, where they now appear perfectly appropriate. Living foliage mingles happily with the "pastoral-sprigged" rococo, and the peasants have made these things their own. Their houses, two-storied as soon as one is north of the Tagus, have inherited the pagoda roofs and curving finials of decayed mansions, and the palms and araucarias of abandoned gardens. They water their flocks or fill their jars at coquettish little fountains ornamented with pink and white scrolls and spirals.

These peasants are called *saloios*, a name derived from the Arabic *çalliao* (the tribute paid in Lisbon by the Moorish bakers). Clad mostly in light grey, with broad open faces, the men one and all wear the tasselled stocking cap of black wool called *gorra* or *barrete*.

The hills follow the northern bank of the Tagus as far as Santarem. On the other side of the river lie the low flats of the Ribatejo, the centre of the bull-breeding industry and of the tauromachic cult. The passion of the people of this region for bulls is reflected in an amusing anecdote. A dumb Ribatejano, so the story goes, presented himself at the gates of Paradise and sought admittance. Since, owing to his infirmity, he could not state his place of origin, St Peter, unable to complete the necessary registration form, refused to let him in. At that moment, a bull happened to pass (for no Ribatejano could conceive of a paradise without bulls), and the dumb man flung up his arms with a cry of "Hé! Touro!" "Ah, then you're from the Ribatejo," said St Peter, "come along in."

The Portuguese bull-fight is a very different affair from the Spanish. The bull's horns are sheathed. The horses of the *cavaleiro* are beautiful animals, and it is considered a disgrace if their flanks are even lightly brushed by the bull's horns. The bull is not killed, and, though there are people who say that they do not like to see him turned into an animated pin-cushion, he appears as oblivious of the pin-pricks to which he is subjected as a footballer of a hack on the shins in the heat of the game.

Gay in the satin coat, lace ruffles and three-cornered hat of eighteenth-century court dress, the mounted *cavaleiro* awaits the bull's onslaught with a *ferro*, or light wooden lance, ornamented with coloured paper, upright in his hand. A trumpet blares. A trapdoor is flung open, and the bull rushes out. Manœuvring him into position close to the barrier the rider wheels through the gap. The bull swerves in charging, his horns a hair's breadth from the horse's flank. With a sharp backward stab the *cavaleiro* plants the *ferro* so that its head snaps off in the tough sinews of the bull's neck. Then he flashes away with a defiant caracol. The skill and dash of the horsemanship and the graceful elegance of this method of fighting make it a joy to behold. When several of these darts have been planted another trumpet call resounds, a herd of cows with tinkling bells are driven round the arena and the bull good-humouredly follows them out again, thus surviving to "fight another day".

More attractive even than the *tourada* are the activities connected with bull-breeding which one may see in the broad pastures of the Ribatejo. The bulls are tended by men called *campinos*, to whom a life in the saddle has given the stocky figure and bow legs of jockeys. They wear the *verdegaio* or green and red stocking cap, with white shirts, scarlet waistcoats, black knee breeches and white knitted stockings. With their brass-inlaid box stirrups and long staves used to guide the bulls it is a thrilling sight to see them galloping across the green pastures.

Before a bull-fight at Vila Franca da Xira or Santarem, or on the occasion of a *ferra* or *tenta* when the young bulls are branded or have their mettle tested, the beasts are driven through the crowded streets of the town. From the security of a first floor window I once saw an *espera de touros* at Vila Franca. Throwing caution to the winds, the whole population stay in the streets to show their valour. Shouts and thudding hoof beats are heard from the outskirts of the town. A moment later, two or three *campinos* thunder past followed by a little knot of bulls and then more *campinos*. The crowd scream and hoot deliriously. But all is not yet over. One of the bulls has become *tresmalhado*, that is to say, separated from the rest. He comes into view at the end of the street, turning every now and then to face the valiants who with coats or towels provoke him. He is in a dangerous mood. Under the very balcony where we stand he cuts in between two drawn-up cars and, to our horror, traps an unsuspecting chauffeur against the wall. Fortunately the man is caught between the sharp horns, not on them, and, as the bull moves on, he slithers to the ground, badly scared but unhurt. Rumours pass from mouth to mouth among the crowd. The bull has broken a man's leg in the next street. Another is being rushed off to hospital. A little farther down the street something in the door of an inn irritates the bull, and he trots resolutely through the doorway. In the twinkling of an eye a score of people are spilt out of the windows. We hear cries, enraged bellowings, and the stamping of hooves, but can see nothing. Presently he emerges and disappears down the street. When all is clear, we crowd into the inn and view the damage—a marble-topped counter shivered into fragments. People are saying that the bull tried to chase a woman upstairs.

A broiling day in July. A *ferra* at Muge. A hundred yearling bulls have been brought in to be branded with the crown and

C of the ducal house of Cadaval. With a pallisade of carts, hurdles and benches a square space has been enclosed into which the little bulls are driven two or three at a time. The young men of the village, who are there for their sport, provoke them playfully. There are many spills, but no one seems any the worse and the whole proceedings are conducted in a spirit of wild hilarity. Presently someone takes a *pega*: that is to say, he stands in front of the bull and, as the latter charges, seizes the two horns in a firm grip and is swung into the air. Others close in, grasping the bull by the neck, flanks and tail, and bringing him to a standstill. A noose is thrown round his legs and he is pulled to the ground. Unless, as often happens, another bull still at liberty scatters the little group with an unexpected charge, the tethered beast is quickly branded, vaccinated, set on its feet and pushed or dragged from the ring.

Beyond Santarem, where the Zezere emerges from a welter of pine-clad hills and gorges, stands Tomar, headquarters of the Order of Christ. Apart from the beauty of its situation, the castle-crowned hill overhanging the little white town and looking across to a tenderly radiant hillside of olives and cypresses; apart too from the eight cloisters of the Convent of Christ and the octagonal Church of the Knights Templar, a pilgrimage to Tomar must be made by all who would under-stand Manoeline Art, Portugal's most original contribution to architecture.

Appreciation of the Manoeline style is largely a matter of taste. There are many who find it too elaborate and ornate. By others it is admired for its evident sincerity and its historical appropriateness. It burst upon Portugal in the age of dis-covery, and the very violence of its explosion proved its undoing. The rapidity of its development is well illustrated by two churches in Setubal, the Church of Jesus, built in 1494, and described by Hans Andersen as "the most beautiful little church I ever saw" and that of São Julião, the Manoeline features of which date from only nineteen years later. Yet in these nineteen years the flamboyant Gothic lines of the first have been overgrown and stifled by the florid growths, as of some tropical creeper, into which Manoeline Art so soon degenerated.

It seems as though the arts in Portugal developed in an environment which brought them too quickly to maturity. They are like fruit which had begun to rot before it was fully ripe. Portuguese literature, for instance, seems to have just missed a Golden Age. Gil Vicente and Camões were respectively the precursor and the successor of a Shakespeare who never came to blend the rich humanity of the one with the formal perfection of the other. In the same way, Manoeline architecture, for all its youthful spontaneity, contains within itself the germs of decay.

Admittedly, the Manoeline style suffered from the fact that it was used principally for adding to existing buildings and loading them with unessential ornament. On the rare occasions on which Manoeline architects were allowed a completely free hand, as in the Jerónimos at Belém, and the west front of the Church of Santa Cruz at Coimbra, they showed themselves capable of creating noble architectonic forms. All the same, it is, in the last event, as a form of decoration rather than as an independent architectural style that Manoeline must be judged. To its exuberant exponents, intoxicated with the glories of their age, everything was grist to their mill: Norman, Gothic and Moorish features; the arabesques and medallions of the Italian Renaissance; the heraldic emblems of Prince Henry the "Navigator," the armillary sphere and the Cross of Christ. Above all they appropriated the naturalistic symbols of maritime discovery—anchors, knotted cables, and, in the rose window at Tomar, the bellying sails of caravels. The cloister of the Jerónimos has almost the corroded, coral-encrusted look of things that have lain for years at the bottom of the sea.

How great a contrast there is between the "broccoli" door near the sacristy of Alcobaça, and the lace-like carving of the great archway in the unfinished chapels of Batalha; between the simple grace of the Torre de Belém and the massive heaviness of the choir window at Tomar. Nevertheless, all these diverse ingredients are unified by the divine spark of inspiration. It is when the long-dead impulse which gave birth

to this style is artificially galvanised to engender the mock-
Manoeline of the Rocio station in Lisbon or the hotel at
Busaco that the result is disastrous.

Beyond Torres Vedras the main road to the north passes
through Obidos, one of the most perfect walled cities in Europe,
like Avila, but, thanks to the cheerful whiteness of its houses,
an Avila in the major key. After Caldas de Rainha come the
two great monasteries of Alcobaça and Batalha: the first a
perfect Cistercian interior in agreeable contrast with the
medley of styles of its outer walls; the second a jewel of pure
Gothic built in a green hollow in the hills to commemorate
the battle of Aljubarrota (1385), at which a few hundred
English bowmen (*gens d'armes anglois si peu qu'il en y avoit*)
helped Nuno Alvares, the "Holy Constable", to rout the
Spanish invader.

The road runs parallel to the sea but some way inland. Only
for a few miles is the long, low coast of Portugal from Minho
to Guadiana graced with a Corniche road. For hundreds of
miles these lonely wastes of dune and unending beaches of
fine white sand, the *areias de Portugal* of the ballad, are enlivened by only a few primitive villages of fisher-folk, many
of whom have preserved uncontaminated the blood of the
Phoenician and other colonists who settled there over 2000
years ago. The very name of
the Phoenicians survives in the
modern Peniche, and their blood
still runs in the *poveiros* of Póvoa
de Varzim and the *varinos* of
Ovar and Aveiro, who have in
their turn founded settlements
up and down the coast.

At Apúlia (Minho), the unmistakably Roman name of
which is echoed in the Roman
noses of its inhabitants, the *sar-*
gaceiros or seaweed gatherers rake in the sea dulse in July
wearing a stiff sou'wester like a Roman helmet and a white

woollen coat with a broad leather belt which reproduces with astonishing fidelity the skirted tunic of the Roman legionary. Of Phoenician origin are in all probability the wonderful crescent-shaped boats in which the fishermen of Caparica, Furadouro and Espinho fish for sardines. Like the Tagus barges, the stem of these boats is lifted upwards in a springing curve to face the cruel surf and actually crosses at the prow the double line of the bulwarks. Those of Caparica, white and blue, rose or coffee coloured, are painted on the bows with a pair of enormous eyes. On the even larger craft of Furadouro, rowed like galleys with four or five men to each oar, the eyes are replaced by paintings of St Anthony or the Miracle of Fátima.

Perhaps the most interesting of the Portuguese fishing villages is Nazaré near Alcobaça, its white streets sheltered by a high cliff on which stands a famous shrine of Our Lady of Nazareth. On the shore which stretches away into the illimitable distance of the south, the fishermen may be seen launching their stocky surf boats, dragging them up the beach with oxen, or hauling in their seines. The men wear the black stocking cap and shirts and trousers of enormous checks which give them a comic clown-like air. The women have black pork-pie hats with big pom-poms at the side, and black hoods drawn so closely round their mouths as to give them the appearance of being veiled.

A no less curious form of headgear is worn by the women of Leiria which lies inland some thirty miles to the north. Under a precipitous crag, crowned by the castle of King Dinis, they hold their market, wearing the little velvet hats of pill-box shape with a wisp of feather laid across the crown, which our great-grandmothers affected early in Queen Victoria's reign. The surroundings of Leiria, laved by the willow-bordered waters of the Liz, are green and sylvan. Beyond, the road winds across the sandy, pine-clad heaths which form the most characteristic landscape of this part of Portugal.

For a few miles along this road the astonished motorist will see the children fall to their knees with folded hands as though in adoration of his machine. Actually, they are begging for ha'pence. But tradition offers a more picturesque, if not too

reliable, explanation of this local custom. When Pedro I
crowned his dead mistress Inez Queen of Portugal and caused
her coffin to be borne in state from Lisbon to Coimbra, he
ordered the peasants to do obeisance as, folded in death, she
passed along that road. It is said to be in memory of this behest
that the children bow the knee to-day to every passing vehicle.

At Pombal, round hats are replaced by orange 'kerchiefs
worn with white bodices and black skirts. In the little castle
town, and in two neighbouring villages named Abiul and
Avelar, an extraordinary custom was associated with one of
the religious festivals of the summer until, quite recently, it
was prohibited by the bishop. On the eve of the festival an
oven was heated into which, on the day itself, a man entered
with a carnation in his mouth and an unbaked cake in his hand.
Walking three times round the inside of the oven, he emerged,
the cake baked, himself unscathed like Daniel in the fiery
furnace. This fakir-like performance was perfectly genuine, but
the secret of it was carefully preserved and has never been
divulged.

A few miles beyond Pombal, Estremadura marches with
Beira.

CHAPTER II

THE NORTH

There is a quality peculiar to university towns which other ancient cities do not share. It is as though, in the flower of their age, they had stumbled upon the secret of eternal youth. *Si jeunesse savait, si vieillesse pouvait*, runs the old proverb. But university cities have succeeded in harnessing the vigour of youth to the wisdom of maturity. They are the slowly formed product of the vitality, the enthusiasms, the illusions of each succeeding generation, superimposed upon the humanism and philosophy of those that have gone before.

In Coimbra one is intensely aware of this spirit. It is the third city in Portugal, was once the capital, and is still the seat of the second-oldest university in Europe. The white houses spilt over a hill above the River Mondego impart to it an air of youthful charm and freshness. Both materially and spiritually it is dominated by the tall clock tower of the university, a shining beacon to which the eyes of the city seem eternally to be lifted. In the narrow, winding streets one is constantly meeting bare-headed students, in black frock coats and flowing gowns, carrying satchels tied with ribbons in the different colours of the faculties: yellow for medicine, red for law, blue for letters and so forth.

The university buildings are in that late Italian Renaissance style which seems most appropriate to a seat of learning, and which attains its most refined expression in the graceful loggia called the *Via latina* and the panelled ceiling, grey and gold alternating with grey and silver, of the *Sala dos Capelos* (Senate House).

University life, here, is in some ways similar to and in others different from our own. An elaborate code of traditions and gradually earned privileges makes the life of the *caloiro* (freshman) scarcely worth living. The students live in hostels known as *repúblicas*, and indulge in "rags" culminating in the *Queima das Fitas*, when, on the conclusion of their fourth year, the *quartanistas* make a public bonfire of their ribbons. Some days after this ceremony the *República dos Grilos*, painted with two enormous cockroaches and the mysterious inscription *Ab Urbe Condita A.D.* 70, was still festively hung with a guy, a couple of wicker chairs, a bath, a guitar and a variety of household utensils, some of the most intimate nature. This building belonged at one time to the professors, but some thirty years ago the students, acting on a sudden glorious inspiration, ejected them, and installing themselves in their place have never since been dislodged. Hero-worship at Coimbra is not dependent on athletic prowess, but is bestowed on those who sound most insistently the "romantic" note, in song, poetry and extravagant actions.

Once, at about two in the morning, I was awakened by the thrumming of a guitar in the street below. The terraced city shone palely in the light of the moon. In a deep shadow on the opposite side of the street four or five students were leaning against a wall looking up at a window. Presently one of them began to sing in a warm tenor voice one of the old *fados de Coimbra*, full of melancholy yearning. To me, as doubtless to many others, music is often the golden key which, opening wide the doors of sensation, leads to apprehension of the secret things which lie in the hearts of places and of those who dwell in them. So it was with Coimbra. That song, floating upwards through the night, revealed the spirit of Coimbra, as only its sons know and love it.

The romantic atmosphere of the place is enhanced by its idyllic setting. It is hard to believe that there can be a greener spot in Portugal, even in the verdant Minho. From the terrace of the university the scene is unfolded in all its loveliness. The Mondego, the only river which is Portuguese from its source to the sea, flows slowly down the wide valley, between lush meadows, fringed with willow, sedge and poplar.

In the background are the blue mountains of Lousã. The river is so shallow that low sandbanks are uncovered in mid-stream, where women wash their linen and men with ox carts dig out the tawny sand. The *tricanas*, as the women of this region are called, are very handsome, and their shawls, flung bandoleer-wise over one shoulder in imitation of the students' gowns, give them a great air of elegance.

Below Coimbra the river passes the *Choupal*, a dense grove of enormous poplars, eucalyptus, acacia and sycamore with an impenetrable undergrowth of willow, alder and tall grasses.

Square-sailed boats drift down from the upper gorges of the Mondego, bringing timber and brushwood. They are long and narrow, like native canoes, with pointed prow and stern curling six feet into the air. The tiller is controlled by two long cords carried up to the bows, so that the course may be changed quickly from any part of the boat. When there is an adverse wind, or none at all, they are punted or "quanted" in Norfolk Broads' fashion.

There is much to see in Coimbra besides the university: the Romanesque *Sé Velha* or "Old Cathedral" and two or three other pre-Gothic churches; the most attractively housed museum that I know; the long monastery of Santa Clara where the "Holy Queen" Isabel, for whom loaves were miraculously turned to roses, lies incorrupt in her silver coffin; and the Quinta das Lagrimas, the scented garden where one of the great love stories of the world, that of Pedro I and Inez de Castro, was played out to its tragic conclusion.

Coimbra was formerly the capital of the province of Douro, now incorporated with that of Beira, the name of which, meaning "border", was given to the country which marched with the ancient county of Oporto. Beira was already divided into two parts, Beira Alta and Beira Baixa, and for the sake of convenience the old Douro province is sometimes called Beira Mar. It extends northwards to Oporto, a country of fields, vines and pine-clad heaths, separated from the highlands of Beira Alta by the Serras of Gralheira and Caramulo, which, in their turn, are cleft by the lovely Val de Vouga. After Coimbra the most important town is Aveiro, the "Venice of Portugal", built on the inextricable tangle of canals, dykes and muddy flats of a lagoon which stretches north to Ovar. Here the pork-pie hat reappears, suggesting that it is connected in some way with the sea, far from which it is never found. Along the water-ways the ancient sharp-prowed *barcos saveiros* and *moliceiros* steal softly along, carrying reeds, seaweed or salt from the white pyramids which glisten on the near horizon like remote snow mountains.

Busaco, one of Portugal's most famous resorts, stands just in Beira Alta, half-way up the ridge where Wellington routed the French. The little monastery, with its queer mosaic incrustations of charcoal, stands in the midst of a humid forest of cedars, oaks and exotic trees which is as great a marvel as Cintra. From the Portas de Coimbra the *landes* and sandy coast of Beira Mar stretch as far as the eye can reach. At the back, from the Portas de Viseu, a vast panorama of the pines and rocks of Beira Alta extends from the Caramulo to the Estrela.

To me the name of Beira Alta will always conjure up memories of Canas de Senhorim and of dawn rides and moonlight picnics from Charles Harbord's enchanting little Hotel Urgeiriça. In the brilliant sunshine there is a hard glitter as of enamel in these unending pine forests cut by the deep rocky gorges of the Dão and Mondego. One can ride for hours in solitude along soft needle-strewn paths, through a country which might be the scene of childhood's fairy-tales. These kindly peasants who ply their axes among the trees, are they not the very wood-cutters and charcoal-burners whose children we followed through so many breathless adventures; that gnarled old woman, bent double beneath a load of brushwood, the witch into whose clutches they fell; and that cottage in the clearing, a wisp of blue smoke curling from its roof, the original

Knusperhäuschen, its huge granite blocks lumps of chocolate and sugar-candy?

Every now and again the forest opens into a clearing where, as in a mediaeval Book of Hours, one may follow the seasons in the primitive round of "works and days". Among patches of maize and vine the granite floor shows through in rounded, glacier-polished boulders against which the grey cottages lean for support, and in smooth slabs of rock on which the peasants thresh or spread their golden maize to dry. Oxen turn a creaking water wheel, or drag heavy, lumbering carts. Yellow pumpkins sun themselves among the stubble. At vintage time the grapes are tipped into tall casks where they are crushed by the feet of treading children, and when it is over the vines which were dusty grey in spring and shrill green in summer expire in a flaming sunset of ochre, vermilion, copper and crimson lake. Over all, remote and diaphanous, hovers the blue shadow of the Serra da Estrela.

Less than twenty miles away, the cathedral of Viseu faces across an open square one of those black and white baroque churches with which Portugal is sown from end to end. Viseu is the capital of Beira Alta and must once have been very lovely. Here the legendary Viriato made heroic war on the Romans, and the no less legendary Grão Vasco (since it cannot be determined whether he was a man or a school) painted in the sixteenth century his middle-aged Virgin Mary and his haunted and haunting St Peter.

From Viseu and Mangualde, Beira Alta surges up to the Spanish frontier and the heights above the Douro. The pines thin out and are replaced by clumps of Spanish chestnut, and the boulders are scattered in ever greater profusion. Roads lead northwards through Castro Daire or Moimenta da Beira, over bleak plateaux where granite villages, built without plaster or whitewash, huddle together like sheep; and where the rugged shepherds, in their cloaks and leggings of inter-woven reeds, have the air of straw-encased wine bottles. In June, like a village bride, these bare highlands are decked out in the gold and silver of low-growing broom. Up near the Spanish frontier, one comes upon little fortified hill towns,

Aguiar da Beira, Trancoso, its ring of battlements still unbroken, Pinhel, Almeida and Figueira de Castelo Rodrigo, glaring across the frontier at that other namesake of Roderic the Goth, Ciudad Rodrigo.

Of these fortress towns the most important is Guarda, standing over three thousand feet above sea-level, on a spur of the Serra da Estrela. It is traditionally known as the city of the four F's: *fria, feia, forte e farta* ("cold, ugly, strong and well-provisioned"), though the second epithet is belied by the beauty of its cathedral in which, more than in any other Portuguese church, successive styles of architecture have been blended into a harmonious whole. With amusement one notices how the gargoyles running round the roof on the north, south and west sides, change, over the east end, to stone cannon mouths pointed menacingly into Spain.

Guarda is the capital of Beira Baixa. It is hard to know in what way this Beira is "lower" than the other, for it is if anything higher in altitude and stretches as far to the north and no farther to the south. It is a thin strip of treeless upland between the Douro and the Tagus. After Guarda, its most important city is Castelo Branco, a white town (though the castle is tawny brown) on a hill commanding a wide view of plain and mountain. Here, as in the Alentejan *charnecas*, we are far from the "garden of Europe". But this landscape with its rolling expanses of brown hill and blue mountain has an austere and compelling majesty.

Down a steep flank of the Estrela, Covilhã spills the cataract of its white houses. Opposite, across a broad and shallow valley, a heaving sea of rose-coloured hills piles itself up on the Spanish frontier, tossing on its unbroken crests the ancient, forgotten cities of Sabugal and Penamacôr. It is a desolate country, thinly populated. Wearing brown homespuns and soft, black hats, the peasants have the motionless, statuesque pose, the weatherbeaten faces, the vague, unseeing eyes of Zuloaga's Castilians. The only touch of colour in this sombre landscape is the red, black, yellow, green and white stripes of the Guarda rugs which the men carry folded over one shoulder.

We found these blankets on sale at the great *Feira dos Santos* (All Saints' Fair) at Pinhel on November 1st. The little square with its fortress tower and its baroque town hall was gay with them. Besides the usual tawdry knick-knacks, the booths were full of blue-and-white saddle-bags, *coroças* (reed cloaks) at the equivalent of a shilling, distaffs made of splayed-out canes, and coloured pottery reflecting in its brown and yellow glaze the slanting rays of the sun. There were ox carts, with pointed prows like boats; and flocks of sheep, their heads bound round, like Easter eggs, with coloured ribbons and tassels springing from their horns.

Between the two Beiras, the Serra da Estrela interposes its bulky massif, no longer two-dimensional as seen from Urgeiriça, but solid and substantial, with long valleys disappearing among its folds, leading to villages hidden and undivined. Up one such valley runs the road from Belmonte to Manteigas. It climbs almost imperceptibly beside a shallow river, which tumbles over its bed of white shingle clear and translucent as a Pyrenean *gave*. In October, on the lower slopes, the poplars and chestnuts had turned to crimson and the larches to a sere yellow. Above, the bare hillside was a dull bronze. Only the pines retained their unchanging green.

For so remote a spot Manteigas is (to the seeker after atmosphere) disappointingly infected with the virus of modernity, with the spirit that has engendered municipal gardens and bandstands, and those useful but prosaic organisations the *Filarmonica* and the *Bombeiros Voluntários* (brass band and voluntary fire brigade). But as he rounds, one after another, the hairpin bends which lift the road swiftly to the heights, the traveller sheds the burdensome mantle of the centuries and is transported back into the bucolic world of Gil Vicente. The flat stretches on the skyline, of turf, scrub and jutting rock, are given over to browsing flocks, watched by shepherds who might have walked straight out of the *Tragicomedia Pastoril* Leaning heavily on long sticks, they wear brown cloaks, broad hats or peaked hoods, and drape themselves in the blankets of Covilhã or Gouveia, patterned in brown and white. Up here it is still easy to imagine Manteigas laying before the Queen

Plate I

THE *CAPA DE HONRAS* AT
MIRANDA DO DOURO

A FISHERMAN AT NAZARÉ

Plate II

LAUNCHING AT NAZARÉ

an offering of "milk enough for fourteen years", Gouveia
"two thousand sacks of chestnuts", and Ceia "five hundred
fresh cheeses, all made overnight, and, further, three hundred
calves, a thousand sheep and two hundred lambs such that fatter
could not be found on any hill". At any moment one expects
these shepherds, like Lopo in the *Tragicomedia Pastoril*, to
break into song and dance "in the manner of those of the
mountains".

Now the hillside grows more desolate, and the rocks are
scattered in wilder profusion. As the traveller approaches the
summit, it is as though he were remounting ever farther the
descending stream of time, until at last he finds himself back
in the chaotic infancy of the world. In such a setting, it is easy
to imagine how the great, primitive nature cults grew up, and
how, among these contorted crags and sombre tarns,[1] in the
unplumbed depths of which the peasants hear the echoes of
distant ocean storms, mankind discerned unrelenting intentions
in the random play of the elements.

Side by side the three Beiras run northwards to greet, across
the turgid waters of the Douro, the provinces of Minho and
Trás-os-Montes. According to a popular legend, Guadiana,
Tagus and Douro, the three great rivers of Portugal, slept
one night where they rise in the heart of Castile, having agreed
to run a race to the sea next morning. Guadiana was the first
awake. Choosing the smoothest path, she set off in leisurely
fashion, meandering through the southern lowlands. Tagus,
waking next, sought to make up time by taking a straighter
if rougher course to the sea. Lastly, Douro woke to find herself
alone. In panic haste she set off for the nearest point on the
coast, heedless of obstacles, carving her way through the rocks
which hemmed her in. Thus do the folk account for the green
margins of the Guadiana, the brown banks of the Tagus and the
precipitous gorges through which the Douro swirls to the sea.

---

[1] In one of these, the Lagoa Escura, legend places a buried palace in which
is guarded a king's cloak covered with diamonds and worth the price of seven
cities. Anyone who wishes to enter must take a black nanny-goat across the
lake and wait till noon when he will find a fig tree which marks the only entrance.

For perhaps a quarter of its westerly course, from Pesqueira down to Regoa, the Douro washes the *país do vinho*, the district, so much smaller than one had imagined, where all the port wine of the world is produced, and where, with a start, one recognises, in huge black letters on white vineyard walls, names long familiar from wine labels or hoardings. On the pale steep hills over the river the terraced vines form a darker shading. Early

in October, for a fortnight or three weeks, the vintage wakes the valley to unwonted life. All day, men and women toil in the vineyards, plucking the dark grapes and carrying them down to the press. With baskets on their backs, they march in single file, the leader blowing rhythmically on a whistle to mark the step. All night they tread out the ice-cold must. The great event is the "breaking" of the first *lagar*. The bruised grapes are heaped up in shallow cement tanks. With guitars or accordions the treaders line up knee-deep on either side and advance to the middle. To keep up their spirits they sing, dance and shout unceasingly. To the honoured guest a drop of the spurting liquid is offered on a bare, upturned heel. To his relief he learns that it may be declined without giving offence. After the treading, for twenty-four hours or more, the must is left for the sweetness to ferment out of it. When the desired strength is reached, it is poured into huge casks, already one-sixth full of raw brandy. A few months later, it is taken down by ox cart or river boat to the wine lodges at Vila Nova da Gaia, where time and treatment change the thick cloudy liquid into the ruby or tawny wine that we like so well.

During our first year in Portugal one province called to us beyond all others. This was Trás-os-Montes. As the name implies, it lies behind the unbroken wall of mountains running from the Galician frontier to where the Douro is overhung by the great bulwark of the Marão. This mountain is the key of

Trás-os-Montes, which to us, as to Camilo Castelo Branco, appeared an "almost fabulous province", shrouded in mystery. To most Portuguese it still remains *terra ignota*, and Trás-os-Montes itself has little concern with the outside world, retaining a conservative, patriarchal mode of life and a spirit of sturdy independence well reflected in the saying: *Para cá do Marão, mandam os que cá estão* ("On this side of the Marão their word is law who dwell here").

Vila Real is the capital, the *Vila Real alegre* of the song, a cheerful white town steeply terraced above a gorge against the dark backcloth of the Marão. The low houses with wooden balconies and great stone escutcheons reflect an aristocratic leisure remote from the hustle of modern life. It is, indeed, in spirit rather than in setting that Vila Real is most *trasmontana*. The wine country and the wooded road running north to Chaves are not the typical scenery of the province, which can scarcely be said to begin before Murça das Panoias. Accordingly, as soon as might be, we set our faces to the north-east, pilgrims to Ultima Thule.

Murça is famous chiefly for its pig which stands on a pedestal in the middle of the village. This, if pig it be, is one of those prehistoric monsters roughly fashioned in stone which are found in various parts of the Peninsula. It is a tradition that whenever elections are held the pig is found next morning daubed with the colours of the winning party. Its name has passed into the language, and to call a man *porca de Murça* is to call him a political turncoat.

Opposite the pig stands the pillory, symbol, not of the seesaw of party politics, but of ancient municipal liberties. All over the country, but chiefly in the north and east, these sculptured pillars survive, often in quite small villages. For Portugal was never feudal. From the days of the Reconquest, the liberty and independence of her townships were assured by special charters, some of which, as at Viana do Castelo, went so far as to deprive nobles of the right to live within the municipal bounds. The pillories served a threefold purpose. First and foremost their object was practical, since felons were made fast to them and there deprived of hands or ears. Secondly, standing generally before the municipal building, they constituted the emblem as well as the instrument of municipal jurisdiction. Finally, they were also decorative in intention, and to-day reflect in their infinite variety the many phases through which Portuguese art has passed.

Beyond Murça lies Mirandela. Between the two one enters the *terra fria* or cold land (in contradistinction to the *terra quente* lying farther East near Moncorvo).

Somehow or other we had expected a rugged, stony land-scape with grey screes and narrow defiles. Instead, as the Marão paled behind us, we found ourselves crossing a bleak, treeless plateau stretching away to low, lumpy hills on the horizon. This is the real Trás-os-Montes where, in the words of Luis de Almeida Braga, "the wind and the light cry to Heaven the desolation of the earth". So high is the mean level of the land that mountain ranges which on the map show the imposing height of four or five thousand feet, dwindle, when approached, into rolling downs.

As the road wound over barren heaths towards Braganza, the serras of Bornes and Nogueira flattened out and were merged into the undulating upland horizon. It is indeed a land of topsy-turvy in which the mountains appear less to rise from 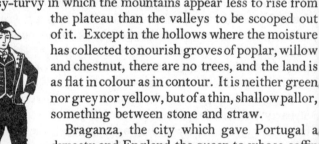 the plateau than the valleys to be scooped out of it. Except in the hollows where the moisture has collected to nourish groves of poplar, willow and chestnut, there are no trees, and the land is as flat in colour as in contour. It is neither green nor grey nor yellow, but of a thin, shallow pallor, something between stone and straw.

Braganza, the city which gave Portugal a dynasty and England the queen to whose coffin in São Vicente our tourists pin their cards, is chiefly remarkable for its citadel and its crypto-Jews. The chance discovery of these last scarcely twenty years ago is one of the most romantic episodes in the history of Israel. When, at the end of the sixteenth century, the Sephardim Jews were expelled from the Peninsula, many were forcibly baptised and suffered to remain. They became known by the engaging name of *Marranos* (swine). It was to ensure that they should not fall away from orthodoxy that the Inquisition was primarily set up, and, by the middle of the eighteenth century, the Holy Office was under the impression that it had completed its task. Nevertheless, many continued to practise in secret the ceremonies of Passover and Atonement, and in order to escape detection settled in the remote, unpopulated regions round Braganza, Vimioso and Mogadouro. Continuing, even after the abandonment of persecution, to observe all the externals of Christianity, they gradually forgot not only all but a few words of their language, but also the significance of their rites which they came to observe in the spirit of a secret society rather than of a religious faith. They ignored the existence of other Jews, just as these ignored theirs. The story of their discovery and of the reawakening of their religious consciousness has been told in full by Mr Cecil Roth.

Braganza stands in the far north-eastern corner of Portugal, but there was another city even more remote which beckoned to us. This was Miranda do Douro. To attain it we wound for a whole day over maquis-covered moors past lonely, huddled villages ever mounting until at last the horizon sank to a dead level and we looked across on equal terms at the Castilian tableland.

The frontier here is a natural one. Out of the level, unbroken rock the Douro has gouged a deep canyon across which, it is said, Wellington was slung in a basket. On the very brink of the abyss, dreaming of its past glories while it crumbles and disintegrates in the sun and the frost, stands Miranda do Douro. Once a frontier fortress of importance, it has declined to a village of six hundred souls living in fifteenth-century houses between the shell of a castle and a great Renaissance cathedral which has long since been deprived of its bishop.

Se fores a Miranda vê a Sé e anda ("If you go to Miranda see the cathedral and depart") runs a local proverb. We stayed the night and regretted it. "Do you remember the inn at Miranda?" we shall always ask ourselves, in parody of Hilaire Belloc. It was one of those places where in Murray's subtle phrase one may "pass the night but not sleep". Next morning we were eager to be away. There was a shorter road back to Braganza, but the car was a "Baby", and the track, we were assured, was "For Adults Only", cars, that is to say, of "Over Sixteen" horse-power. But we took the risk and were justified in the event, dodging the boulders and tilting to one side or the other until it seemed we must turn turtle. In the bright morning light, the country was its own reward, the rolling heaths a mosaic of gold and silver broom, purple, aromatic wild lavender, and cistus, the huge white blossoms decking every bush like handkerchiefs spread out to dry. Great emerald lizards ran across the path, and golden orioles flashed among the bushes.

In these "insulated regions", as Antero de Figueiredo calls them, the only patois in Portugal is spoken, a dialect half-way between Portuguese and Spanish. The *terras de Miranda* share with Madrid the *nove meses de inverno e três de inferno* ("nine months of winter and three of inferno") of the proverb. But the peasants, primitive though their mode of life, are sturdy, intelligent and, to judge from their fields of rye, more prosperous than many of their countrymen. The Napoleonic homespun costume of the men has almost disappeared, but one may still see the *capa de honras*, a hooded cloak of brown blanket cloth, with cut felt ornamentation weighing over fifty pounds. The women have no special costume, except perhaps the black stockings which seduced maidens used to be compelled to wear, although, apparently, this occasioned them no difficulty in finding a husband. No decline in spirit has put an end to the village feuds of this region which break out in fierce fights with the *fouce roçadoura* (hedging bill) at the *romarias* to which the young men go in bands, fifty or sixty strong, marshalled behind their village bagpiper.

From Braganza a road runs westwards along the Spanish frontier to Chaves. On the right, low, chestnut-clad mountains mark the borders first of Leon and later of Galicia. To the left the vast expanses of the *terra fria* slope away. Desolate in summer, how much more desolate must they be in winter, wolf-haunted under their mantle of snow. There are few villages and

fewer scattered farms. The only signs of habitation are dove-
cots, low round towers sliced off at an angle, and the dwellings
of solitary herdsmen, miniature straw wigwams set up on ox
carts so that they may be trundled from one pasture to another.

Chaves is remarkable rather for the beauty of its situation
than for its monuments. A notable exception, however, is the
many-arched Roman bridge with inscribed pillars to remind
the traveller that he is on the Roman highway from Braga to
Astorga. In June 1932, when we passed that way, the western
part of this road had just been reopened after many centuries
of disuse. From Chaves it climbs steeply until, three thousand
feet above sea-level, it emerges into a landscape strongly
resembling the Yorkshire moors. On both sides, wide expanses
of heather lead away to the flattened tops of invisible mountains. These are the *terras de Barroso*, a district so primitive in
its mode of life and so remote from the ways of man that it has
earned the name of the "Portuguese Andorra". Montalegre
is its capital, a village raised to the dignity of a county town
by its possession of a castle. It stands some ten miles from the
main road, 3600 feet above sea-level, near where Larouco, one
of the highest mountains in Portugal, rears its gaunt summit
like a head without shoulders above the surrounding highlands.
The people of Barroso depend for their living on sheep, cattle
and a scanty crop of potatoes. The women wear short hooded
cloaks of coarse brown wool, and the men silk 'kerchiefs
knotted round their heads like turbans.
    Heedless of the "splendours and miseries" of Barroso the
road unfolds its white ribbon across the moors, joins the head-
waters of the Cavado and then, dropping towards more hos-

pitable regions, affords dramatic views of the Gerez, not the highest but incomparably the most rugged of the Portuguese mountain ranges.

Presently an inconspicuous borderstone is passed, and we have left Trás-os-Montes for the maritime province of the Minho, or, to give it its full dignity, Entre-Minho-e-Douro. All at once a subtle change comes over the landscape. The eye is caught by vines trellised on tall, granite pillars or trained creeper-like over trees and hedges, and by oxen of a new breed with long curved horns and elaborately carved wooden yokes. Only now, by contrast, is it borne in upon the traveller how totally lacking Trás-os-Montes has been in detail. For detail is the keynote of the Minho, and no greater contrast can be imagined than between the sober, austere monotony of the upland province and the infinite variety of the valleys by the sea.

The Minho is the show province of Portugal. Bounded by mountains to the East and two great rivers to the North and South, it is the veritable "garden of Europe planted by the sea" of the poet Ribeiro. Everything is in miniature and has the bright spick-and-span finish of a toy world. Peasant homes are like diminutive country houses; hills scarcely two thousand feet high have the spacious forms and rugged outlines of mountains; and narrow streams flow with the placid dignity of broad rivers. The very fields are a checkboard of infinitesimal holdings, for there are no great landlords or landless peasants, but countless smallholders each cultivating his own fragment of land. Something of this brightly variegated quality has passed into the lives and character of the inhabitants. The *minhotos* are the gayest and most carefree of the Portuguese. The burden of their work is lightened by cheerful song, brightly coloured costume and a never-ending round of festivals. They seem to sing and dance their way through life.

The very heart of this arcadian land is the valley of the Lima from Ponte da Barca to Viana do Castelo and the far-famed hill of Santa Luzia. It is a broad valley, set between stately hills. Round the grey stone villages stretch the vineyards which produce the *vinho verde*[1] peculiar to the province. Hills and valleys alike are of a rich velvety green, refreshed even in summer by soft Atlantic rains. No wonder that the Romans placed the Elysian fields here and saw in the Lima a second Lethe.

It is at Viana that one may see the richest regional costume of all Portugal. Even on working days there is a distinctive quality in the open boleros and striped black and red skirts of the women. But it is on festival days that the *traje á lavradeira* appears in all its splendour. It comes not from Viana itself but from the surrounding villages of Afife, Carreço, Areosa and others. There are two varieties, the "red" and the much rarer "blue". Except in the latter the predominant tone is a brilliant vermilion.

Over a striped skirt, red and black or red and yellow, with often a broad black border round the hem, hangs a short apron gaily embroidered in a variety of colours. Of the white chemise all but the embroidered sleeves is hidden by a small bolero and a fringed, patterned shawl, the ends of which are tucked in at the waist. Another similar shawl is draped bunchily from the crown of the head, showing the front part of the hair. This brilliant symphony of scarlet, yellow, green and black is completed by mules, white stockings, a handkerchief sachet often in the shape of a heart and all the golden chains, coins and pear-shaped filigree ornaments which the wearer can muster. The whole costume suggests the Balkans rather than Western Europe and relies for its effect less on line than on richness of colour and ornament.

If anything in the Minho is to prove disappointing it will probably be the larger towns. A century ago, these must have

[1] The designation *verde* should in this instance be translated as "unripe" rather than "green". The wine to which it is applied is made from grapes which never fully ripen owing to the heavy rainfall. It has a sharp flavour which makes it peculiarly refreshing, but unfortunately it does not travel.

had great character, but since then they have seen an orgy of indiscriminate destruction. Even Braga, overhung by the twin shrines of Bom Jesus and Sameiro, is only a shadow of the primatial city of all the Spains. Similarly Guimarães, Barcelos, Ponte de Lima, Caminha, Viana and Vila do Conde can each count many lovely buildings but have lost the uniformity of many towns in the south.

On the other hand, nowhere can the national style of domestic architecture be better studied than in the Minho manor houses, many of which have given their names to Portugal's oldest nobility. This style took definite form in the eighteenth century, when Brazilian gold gave an immense impulse to construction. The abundance of limestone and granite, coupled with the relatively poor quality of the timber, nullified the decorative, sculptural influences of the Moors, and, coupled with the classical tendencies of the epoch, bred qualities of solidity and sobriety. The facade of the Portuguese country house is notable for its harmonious proportions and for the symmetrical distribution of the many windows with their more or less ornate baroque enclosures of unwashed stone.

These houses are planned on an ample scale and furnished with imposing entrances. There is often a double stairway, either rising from an internal hall or leading externally to a covered portico. The lack of ornament is counterbalanced by the decorative use, both externally and internally, of *azulejos* (ceramic tiles). Although originally of Arabic invention, and strongly influenced at a later date by the *faience* of Italy and Holland, the Portuguese have made the *azulejo* their own by their extension of its decorative uses. Abandoning first the embossed or mosaic patterns retained in Spain, and later the yellows and greens of the sixteenth and early seventeenth centuries, the *azulejo* came to retain only the blue and white of the eighteenth century which was its apogee. The variety of uses to which these tiles are put is astonishing. Low dadoes are composed of *azulejos* each with its own design like those of Delft but with heavier, thicker brush-work. Whole walls not only of private houses and palaces but even of churches are covered with vast compositions depicting classical legends and allegories, baroque scenes of court or country life (rarely local in atmosphere) framed in rococo floral designs, incidents from the Bible, from the lives of the saints or from mediaeval legend and history.[1]

Only in one corner of the Minho is its diminutive charm amplified into grandeur. The mountains of Soajo and Gerez on the north-eastern frontier, nearly as high as the Serra da Estrela, have long fostered a sturdy, independent breed of men. So free were the people of Soajo in the days of the monarchy that for long they paid no taxes to the King beyond an annual tribute, offered with disdainful arrogance, of three greyhounds. Their costume of white woollen cloth, which has almost disappeared to-day, is one of the most striking in the country.

In the Gerez there still survive curious social institutions which amount almost to a primitive form of communism. Traces of similar traditions are found in Barroso and in the *terras de Miranda* where Antero de Figueiredo found "a judge chosen by themselves, exercising a strict control over all cattle and imposing fines payable in wine, which was afterwards consumed in common". In the Gerez, in the village of S. João do Campo, for example, roads, bridges and sheepfolds are built and maintained by the community. All the cattle are pastured by a single householder in turn, whose period of duty lasts for as many days as he owns head of cattle. Each family is responsible for the cultivation of its own land, but voluntary assistance is given to any who may find the task beyond them. Every head of a family is called upon, in turn, to assume the office of *Juiz* and to take decisions affecting the welfare of the community, in collaboration with the *Acordo* (Council) composed of five other householders, whom he summons with a rousing blast on a horn. Together they debate such weighty affairs as wood-cutting, irrigation, and the dates on which the

---

[1] In the refectory of the Jeronimos Monastery at Belém there is a series of pictures depicting the story of Joseph in one of which the coat of many colours is shown as a pair of nether garments.

sowing and the harvest are to be begun and the cattle turned out to pasture. They also settle any disputes which may have arisen, hearing the evidence, assessing damages and inflicting fines. In all these matters the judge's decisions are unswervingly obeyed.

In its south-western extremity the Minho province grows flatter and less varied. The River Ave meanders pleasantly through vineyards, maize fields and pine woods which grow thicker as Oporto is approached.

Though not the capital, Oporto has the distinction of having bestowed on the whole country its ancient name of Portus Cale. Like Lisbon, it is impressive less on account of individual monuments than of its general situation and appearance. Crossing by one of the two lofty bridges over the Douro to the British wine lodges at Vila Nova da Gaia, the traveller is rewarded with a view which has changed but little since Wellington in 1809 forced the passage of the river in the teeth of the French.

On the opposite side of the gorge rises the city with countless miniature gardens inlaid among its red roofs and grey walls. It culminates in two steep hills, crowned the one by a Romanesque cathedral and the other by the tall Clerigos tower. The spacious squares and avenues of modern development are hidden behind this spectacular drop curtain, and the old Oporto of the river bank is as unspoilt at close quarters as from a distance. The predominance of granite lends the city a more sober appearance than Lisbon, but the street life is no less exuberant.

This centres principally round the Cais da Ribeira, the long quay where cargoes are discharged into ox carts adorned with elaborate carved yokes, shaped like a board and standing as much as two feet above the necks of the oxen. The intricate carving of these yokes strongly recalls certain sculptured motifs found among the stone beehive huts of Citânia de Briteiros near Guimarães, which, whether or not it is the Citânia mentioned by Valerius Maximus, is undoubtedly an important pre-Roman settlement. Nevertheless, the dates inscribed on these yokes are seldom more than ten years old and they constitute one of the most gratifyingly vigorous branches of Portuguese, indeed of European, folk-art.

It is at the Ribeira quay among its sturdy boatmen, strident fish-wives and laughing children that we may most fittingly bring our pilgrimage round Portugal to an end, before examining in greater detail the customs and traditions of its folk. Here, better than in palace, church or peasant hut, better than in green pine wood or flowering *charneca*, we can lay our hand on the throbbing pulse of the race, share their pleasures and their sorrows, feel the measure of their exuberance and their pathos, and appreciate the courage with which they face the countless problems of existence. Here, above all, we may apprehend that humane, because so intensely human, spirit which is the hall-mark of the Portuguese outlook and the richest contribution of the race to the philosophy of mankind.

Part Two

TRADITIONAL BELIEFS AND CUSTOMS

"Iberia's fields with rich and genuine ore
Of ancient manners woo the traveller's eye."
MICKLE, *Almada Hill* (1780)

Plate III

ON THE DOURO

[Casa de Portugal

[facing p. 48

Plate IV

THE PIG OF MURÇA

YOKED OXEN AT OPORTO

CHAPTER III

MAGIC AND SUPERSTITION

Magic is a much maligned art. This is principally because it is a much misunderstood one. Properly, indeed, it should be called not an art but a science, or more exactly the forerunner of science and its "bastard sister".

With science, magic has this in common, that it is founded upon the belief in a series of natural laws by which the sequence of events is inexorably governed, without the intervention of any supernatural or spiritual agency. If magical rites are now regarded as "experiments which have failed", and if magic itself has become discredited, this is not because the existence of any flaw has been established in the fundamental hypothesis, but because the latter was faultily applied at a time when man, with his great lack, both of experience itself and of the opportunity for correlating its results, formed a mistaken conception of the nature of these laws. An association of ideas which it is easy to follow, if not to accept, led mankind to place their faith in apparent relations between cause and effect which, upon more accurate observation, are found to have no existence in fact.

The whole precept and practice of magic rests upon two false assumptions: the first, that things which resemble one another are in effect the same; the second, that things which have once been in contact with each other remain, in effect, in contact. The most obvious and familiar applications of both assumptions are to be found in the attempt to harm an enemy by injuring or destroying either a likeness of him, or some object such as hair, nails or clothing, which has formed part of his body or been in contact with it. The first practice is an example of what may be called "imitative", and the second of "contagious" magic. The first is founded on the association of ideas by similarity, and the second on their association by contiguity. Both may be included in the definition of "sympathetic" magic, since they assume the existence of a secret sympathy or relation between things which are not in actual fact so related.

It must be emphasised that, once these ill-founded premises are accepted, they are applied in practice with the most rigorous logic. The magician seeks to influence the course of events, not by invoking the assistance of any supernatural agency, but by himself harnessing the laws which, to the best of his belief, govern them.

Nevertheless, although magic is thus logically the predecessor of religion, the stage of human development at which the former was practised in a pure and undiluted form to the complete exclusion of the latter must be so remote in time as to lie beyond the bounds of speculation. There are indeed many who will not admit that an age can ever have existed when mankind did not nourish the conception of a superior being with incalculable powers, to whom he could turn when his efforts proved unavailing, and who could, if it so pleased him, make the sun to stand still in the sky.

This much, however, is certain; while the fundamental doctrine of magic is the very antithesis of that of religion, these two irreconcilables have been blended throughout recorded history and the known world in that blundering philosophy by which the greater part of mankind guides its actions. It is because of the eternal conflict in the human heart between self-reliance expressed in magic (and in science) and self-abnegating faith in the inscrutable ways of the deity, that the Church has always been the most implacable enemy of magic, and has to such an extent blackened its reputation.

The fact remains that the conduct of us all is still to some extent influenced by a tangle of precepts and taboos which cannot be explained, still less justified, except by the two fundamental false assumptions on which magic is based. Even if they are not applied with their pristine logic, consistency and faith, these assumptions have engendered countless superstitions, which, as their name implies, are anachronisms inherited by mankind from a former age.

It may well be asked why the futility of magic was not demonstrated long ago by the time-honoured expedient of trial and error. Surely, it may be argued, magic should have been discredited the first time it was put to the proof. Why then has it not long since been abandoned? The answer is not hard to find. Man was ever a credulous creature, and his critical faculty has always been that of which he makes the least use. The sheer momentum of habit will make him continue to believe what he was born and bred to believe, what his king, priest or sorcerer bids him believe, however much his own perceptions, working on the limited data at their disposal, testify to the contrary. Legendary tales of their proven efficacy gather round every magic rite and unite to form a lore of which the savage no more doubts the truth than does the Christian of the New Testament.

More even than the wish, the unquestioning conviction is father to the thought. Far more weight tends to be attached to evidence which confirms than to that which contradicts a conclusion which hitherto has not been regarded as open to doubt. Even if less than half of the rain-making charms which a savage may witness appear to achieve their object, this will be enough to maintain his belief in their general efficacy. Furthermore, rain-making charms, to choose but this one instance, are most usually performed when rain is most urgently needed, most overdue, and accordingly most likely to fall. What wonder, then, if in many cases they appear to achieve the desired result? It need be no matter for surprise, therefore, if in Portugal the old people say, as they do at Santo Tirso, that the crops are less good than in their young days when the *clamores* regularly went the round of the fields.

For the rest, the momentum of acquired habit is among the strongest of human motives. Superstitions continue to be observed, not necessarily with the concrete conviction that they will achieve this desired object or avert that feared evil, but because *é bom* to do so. "It is good", for instance, to ask the new moon for money, to cross one's thumbs in front of one's mouth when one yawns, to place a coin in a dead man's coffin, and to perform a thousand other apparent irrelevancies.

The connection between the superstitious charm or taboo and its original intention has not altogether disappeared in Portugal. In many parts of the country a ruptured child is passed three times through a reed or sapling or sometimes an oak tree, which is split for the purpose. Thereafter, as the plant heals, so should the injury to the child. This rite must be performed at midnight on St John's Eve, by three men of the name of John, while three women, each called Mary, spin, each with her own spindle, on one and the same distaff. The virtue in this rite would be ascribed by the peasants to its religious associations. But primarily it is an application of sympathetic magic, "imitative" in that the healing of the reed cures the rupture of which it is a representation, "contagious" in that the passing of the child through the reed sets up an enduring current of sympathy between them. In order that they may heal together a set dialogue like a spell is exchanged between the Johns and the Marys, part of which runs:

> "What are ye spinning, o ye Marys?"
> "Silken thread
> To heal the reed
> Where passed the ruptured child."

The "imitative" principle is well illustrated by the belief, no doubt translated into practice from time to time, that an enemy may be injured by sewing up with red thread the eyes and mouth of a toad. As the toad slowly perishes, so will the enemy wilt away, whose effigy it is intended to represent.[1]

The same notion lies at the root of a form of divination practised in Beira Alta, to ascertain whether an absent person is in good health. A sprig of a plant called *erva de Nossa Senhora* is cut and placed on the roof. Henceforth this herb

---

[1] In a variant of this belief recorded by Jaime Lopes Dias at Teixoso, the two branches of sympathetic magic are again united. A piece of bread bitten by a person whom it is desired to harm is placed in the mouth of a toad, and a pin run through the latter's head.

represents the person concerned and remains green only as
long as he is well, drying up if he falls ill or dies.

The "contagious" principle, which is, if anything, even
more remote than the "imitative" from modern conceptions
of cause and effect, accounts for a wide variety of superstitious
prejudices against neglect of all severed portions of the body,
or substances derived from it, such as saliva or human milk.
Round Oporto a woman will never throw away her hair
combings without spitting on them and crossing herself thrice.
It is considered dangerous to allow a gnawed piece of bread to
fall to the ground after a meal, because, if some noxious insect
were to consume the infinitesimal quantity of human saliva
which it may contain, the poison would be communicated to
the person who had bitten it. Similarly, the health or character
of a child is thought to be directly affected by the manner in
which the afterbirth is destroyed, and a mother with a child at
breast is careful as to what happens to the least drop of her
milk. A servant employed by some friends of mine was discovered one day to be placing hairs from the tail of the household dog behind the front door, under the impression that she
was thereby curing it of the habit of roaming. So strong a
current of magic sympathy may a single hair exert on the body
from which it was plucked.

Although the sympathetic element in superstition is often
immediately obvious, this is not always the case. While it is
easy to understand the belief that if milk is boiled the udders
of the cows will dry up, it is less clear how burning fig skins
on the fire can similarly affect the milk supply, as is believed
at Belas. There is a certain perverted logic in the idea that the
husband of a woman who gets soaked when washing at the
stream will also get "soaked" in a metaphorical and less
reputable sense. But how may we account for the belief,
recorded at S. Martinho de Bougado, that the first louse caught
on a child's head should be killed in a tin in order that the child
may grow up with a good singing voice? One has heard of
a swan song, but surely a louse expires in silence.

In magic, as it survives to-day, precept and practice do not
always go hand-in-hand. A great variety of seasonal rites,

originally magical ceremonies for the promotion of agriculture, which will be described in a later chapter, continue to be performed, in a more or less modified form, as recreations and games, long after their original purpose has been forgotten. Men will often continue to do what their fathers did before them, while offering a completely different explanation for their actions.

More usually, however, the reverse is the case. The memory of a rite or a taboo, coupled with some vague shreds of belief in its efficacy, lingers in the peasant memory when its practice has long since been abandoned. A case in point is the use of the "bull-roarer" as a children's toy in the Azores. This instrument, identified by Andrew Lang with the *rhombos* used in the ancient Greek mysteries, consists of a piece of wood about eight inches long by three wide, sharpened somewhat at the ends, to one of which a string two or three feet in length is tied. When whirled rapidly round and round, the bull-roarer produces a loud and peculiar whizzing sound. Among certain primitive tribes in New Guinea it is associated with the monsoon and may have once been used for the magical production of rain. Not so in the Azores, which have long outgrown so primitive a stage of belief. In these islands, none the less, the old people still forbid the children the use of this toy at harvest time, on the ground that it is likely to spoil the weather.

It is easy to understand that rites of aversion or charms intended to protect men from dangers, the precise nature or imminence of which they were generally unaware, were as a rule submitted to less searching tests than those which aimed at accomplishing a concrete and difficult task.

Many of the most deep-rooted Portuguese superstitions are those which purport to furnish the credulous with the means of averting a variety of nebulous, ill-defined and often imaginary diseases and disasters. Foremost among these are supernatural malign influences such as *bruxaria* and *feitiçaria* (witchcraft and enchantment). The belief in witchcraft is still very much alive in the country districts of Portugal. On the very outskirts of Lisbon one finds peasants whose fear of witchcraft is such as to make them unwilling even to mention it. The belief differs in none of its essentials from that current in the rest of Western Europe. The witches are regarded as essentially maleficent. It is their delight to blight the crops, to maim cattle or smite them with sickness, to lead the belated traveller astray, and above all to suck the blood of young children so that they sicken and die.

To quote a concrete case, in December 1932, the infant child of Armenio Pereira of Santa Leocadia de Baião died, almost certainly as the result of suffocation through being overlain. The parents, however, were persuaded that the child had been "sucked by witches", and its grandmother went so far as to declare that she had seen the witch flying away in the shape of a black sparrow.

If she suspects that her child has died as the result of witchcraft, a mother will usually breathe no word of it to the neighbours, but will set the child's clothes to boil in a cauldron of water, stabbing them over and over again with some sharp instrument. These stabs will be felt by the witch on her body, and she will be compelled to come to the house to beg for mercy. Alternatively, the mother will take a broom (the emblem of witchcraft) and will sweep her house the wrong way round, that is to say, from the door inwards, saying:

Assim como eu na minha casa ando a varrer	As I go sweeping in my house
Assim quem matou meu menino aqui venha ter.	So may she who slew my child be drawn hither.

The Portuguese witches assume animal forms at will, those of ducks, rats, geese, doves and even ants being more frequently mentioned than the more familiar cats and hares. Their power lasts from midnight till two in the morning, during which time they may be heard clapping their hands, laughing or howling dismally. Although no especial name is given to the Sabbath,[1] the witches meet at cross-roads on Tuesdays and Fridays, thus accounting for the popular prejudice against those days, as expressed in the proverb:

Ás terças e sextas feiras	On Tuesdays and Fridays,
Não cases a filha nem urdas a teia.	Neither marry your daughter nor weave at your loom.

To leave their houses, they strip themselves naked, anoint their bodies with an unguent called *azeite zumbre* which they keep hidden beneath one of the hearth-stones, and disappear up the chimney on pronouncing a magic formula, one of the various versions of which is:

Por cima de silvais	Above the brambles
E por baixo de olivais.	And below the olive trees.

Throughout Portugal, as in the Basque country,[2] the story is told of a man who overheard the formula, but on attempting to pronounce it himself put it the wrong way round, with the result that he was carried underneath the brambles and arrived at the Sabbath in a piteous state.

At their meeting place the witches worship the Devil in human form, or in that of a dog, black cat or black-horned bull calf, not omitting the ritual kiss on the spot sanctified by tradition if by nothing else. Every witch is believed to possess several balls of thread which she must bequeath to a successor before she can die. There is a legend that a certain witch could get no one to accept this ill-omened bequest and remained unreleased from her death agony, until she bethought herself of a big jar which stood near. "I bequeath them to the jar", she cried, and forthwith the jar burst into fragments and the witch breathed her last.

There are several ways of identifying witches. Leite de Vasconcelos was told concerning the prayer of the Guardian Angel that it "has the virtue of stripping witches naked, and

---

[1] Except, in one isolated instance, *Senzala*.
[2] Cf. *A Book of the Basques*, pp. 255–7.

if it is left unfinished they will come to the house of the reciter begging that it may be concluded, so that they may clothe themselves again". This prayer is an enumerative jingle of the "What-is-your-one-oh" type. At Taboa, if, while the Easter procession is on its rounds, you grease with a bit of bacon left over from Carnival the principal door of the church, the witches will reveal themselves, when the time comes for them to leave the church, by appearing unable to find objects, such as a book or a handkerchief, which they have in their very hands. In other places it is held that if the Missal is not shut at the end of Mass, or if a silver coin or a plant or leaf, plucked at the moment when a shooting-star flashed across the sky, be placed in the holy-water stoop, the witches will remain rooted to the spot.

The nature of witchcraft has long been the subject of learned controversy between specialists. In Portugal, less even than in other European countries, does it bear the appearance of having been a real cult of satanism. The most acceptable solution of the problem, put forward among others by Miss M. A. Murray,[1] suggests that the witch cult was no more than the survival in a debased form of a pre-Christian fertility cult which was driven underground by Christianity and endowed with a reputation far worse than it deserved.

One of the principal planks in this theory is the suggestion that the word "devil" or "Satan" was not used by the devotees of the cult, but was placed in their mouths by their persecutors to denote the pagan deity in whom the cult was personified. It is a commonplace that when a new religion is established in any country, the god or gods of the old religion become the devils of the new. The part played by the Devil in Portuguese

[1] M. A. Murray, *The Witch Cult in Western Europe*. Miss Murray falls into the error of trying to make her theory square too completely with all the fantastic details which the over-vivid imagination of the Middle Ages introduced into popular lore and even into the written records of the witchcraft trials. She makes insufficient allowances for the credulity of the period, a credulity which, to some extent, she asks her readers to share. We cannot be asked to believe every statement made by Marco Polo. But there is no doubt that that great traveller really reached China. In the same way Miss Murray's theory is so convincing in its main outlines that it is quite unnecessary that it should account for every detail of a belief which, like every other branch of folk-lore, must have suffered countless mediaeval accretions.

folk-belief affords strong confirmation of this supposition. In the first place, he is known by a variety of names which alone would suggest that he has collected round his own person the characteristics of more than one supernatural being of the old beliefs. The following are some of his aliases: *Diabo*, *Mafarrico*, *Porco-Sujo*, *Bicho negro*, *Galhardo*, *Provinco*, *o da carapuça vermelha* (he of the scarlet cap; cf. the "red Master" of Basque folk-lore), *Trasgo*, *Crespo*, *Manquito*, *Zangão*, *Farrapeiro*, etc. Like the fairies or their foreign equivalents he is credited with the construction of bridges, or the erection of balancing stones. He is sometimes accompanied by a *Diaba*, who is his mother or his wife.

Of the many names by which the Devil is known, by far the most suggestive is that of *Dianho*, recorded by Leite de Vasconcelos at Godim. A ninth-century decree attributed to a General Council of Ancyra states that "certain wicked women, reverting to Satan...believe and profess that they ride at night with Diana". For this and other reasons it is generally believed that the cult which survived in the form of witchcraft was that of the deity known as Janus (Dianus) or Jana (Diana). Although *Dianho* may perhaps be no more than a corruption of *Diabo*, it is not impossible that in this corner alone of Western Europe the Devil of the Christian belief has preserved the very name of the god of the pagan.

In the neighbouring provinces of Northern Spain it is without doubt this name which has survived in various forms such as: *xanus*, *xanas*: fairies (Asturias); *xas*: spirits of a Puckish type (Galicia); *Diañu*: devil (Asturias); *diaño burlón*: faun (Asturias). In Portugal the name of Janas denotes a village near Cintra, of which the church, circular in shape, was clearly built on the foundation of a pagan temple. A Roman inscription found close by suggests that this temple was dedicated to the *Dei Manes*. In the Algarve the memory is not extinct of female creatures called *jãs* or *jans*, for whom it used to be customary to leave a skein of flax and a cake of bread on the hearth. In the morning the flax would have been spun as fine as hair and the cake would have disappeared. If the cake were forgotten the flax would be found to have been burnt during the night. "In the Algarve", writes Leite de Vasconcelos, without apparently realising the significance of the fact, "it is very usual to say *Pobre Jan* or *Pobre Janes* for 'Poor Devil'."[1] It has been suggested that the Latin form of *Janus* may be derived from the root "gen-", in which case it would be hard to imagine a more appropriate name for the deity of a fertility cult.

Other malign influences in which the belief still persists in Portugal are the *mau olhado* (evil eye), *mal de inveja* (overlooking), *ar ruim* (evil vapours), *luada* (influence of the moon), and *quebranto* (general weakness), this last supposedly caused by a look from one of the persons nearest or most dear to the victim. These influences are all held to result in the unaccountable decline of the victim, and having no existence outside the superstitious imagination, can only be countered by means derived from the powers of religion or of magic.

An equally ill-comprehended object of fear is the thunderbolt (*raio* or *corisco*). Even to-day the commonest Portuguese imprecation is: *Raios te partam* ("May thunderbolts cleave you"). In the Azores it is considered unlucky so much as to pronounce the word *corisco* (it being, of course, a common superstition that a person or thing is attracted by the pronouncement of its name). The thunderbolt is conceived in the form of a stone which falls to earth at the moment of the flash, and sinks seven metres deep into the ground. Each successive year it rises a metre, until after seven years it reappears on the surface, where it is seized upon by any peasant who finds it, in the belief that it will protect his house from being struck, since "where one is, no other will fall". Placed on the roof or in the house these *pedras de raio* (or *centelhas*, *faiscas* or *perigos* as they are also called) are believed to jump about in thundery weather when they are exercising their function of amulets. The curious thing is that these are no ordinary stones which the peasants identify as thunderbolts, but prehistoric arrow-heads and other implements including the very emblem of the thunder god, the hammer.

[1] In the *Farsa de Inês Pereira*, Gil Vicente makes Inês exclaim: "Oh Jesu! Que *Jam* das bestas!"

The *pedra de raio* is only one of a large number of different amulets and charms with which the Portuguese peasantry arm themselves against every inconceivable harm, real or imaginary. In one of Lisbon's smartest shopping streets one may buy little bone horns, half-moons and hands making the ancient, bawdy gesture of the "fig". Round the neck of a baby less than a year old in the Serra de Monfurado was hung one such "fig" of polished horn. At the market at Ponte de Lima a cow which had just given birth to a calf wore on its forehead a little bag of red cloth containing salt, rue, garlic and a splinter from a broom handle. A curly ram's horn crowned the steeply curving prow of one of the half-moon shaped sardine boats of Caparica, which already bore gaily painted on the bow the eye affording protection from the *mau olhado*.

The objects which constitute these amulets fall into several classes. Among natural substances thus used are many drawn from the animal kingdom such as toads' legs, snakes' or bats' heads, horns, and bones; from the vegetable world, such as rosemary, wild lavender, broom, garlic, four-leaved clover, nuts with three kernels, and olive stones; and from the mineral, such as iron, steel, jet, coral, stone and rock-crystal. The shapes into which these and other substances are worked may also of themselves carry virtue. There is, for example, the horseshoe, the half-moon, the fish, the heart, the pentacle or Solomon's seal, and the "fig"

Yet another class of amulet is formed by objects to which extraneous circumstances have lent a virtue which would not otherwise reside in them. In this class may be numbered "thunderbolts", herbs that have been blessed, pieces of red thread the exact length of a particular Saint's height, rings from dead men's fingers, fragments of winding-sheets and, at Vinhais, that sinister object the *mão refinada* or "hand of glory", the dried and pickled right hand of a dead man, which is in great demand with thieves since it is reputed to have the gift of opening locked doors and inflicting on householders by sympathetic magic the deafness and insensibility of its defunct owner.

A few further concrete instances of the use of amulets may be of interest. In the district of Elvas, Tomás Pires has recorded a number of examples of which he has first-hand knowledge. Children are defended against *quebranto, luada* and the malefices of witchcraft by wearing a half-moon in silver or copper, a *figa* in silver, marble, coral or jet, a wolf's paw, a perforated coin, or the horn of a goat found unsought on a Tuesday or Friday and twice kicked before being picked up. In teething they will be soothed by wearing a wolf's tooth, and lily roots or garlic are considered (all too wrongly) sufficient to keep them free from vermin. A mother can assure herself of an abundant supply of milk by carrying on her person a key, some object made of jet or a globule of milky tinted agate. Knives are a specific against haemorrhage, and against epilepsy it is thought prudent, in one locality, to carry nails from the shoe of the right forefoot of a horse which has drawn the image of St George in the Corpus Christi procession.

The protection afforded by all these amulets is conceived in the most concrete manner imaginable. At São Martinho de Bougado one may hear of a *figa* which saved its wearer from being struck by the evil eye by attracting to itself, like some lightning conductor, the full virulence of the look, and was thereby split into two pieces. To understand the principle on which amulets are supposed to work (for despite all appearances to the contrary they are governed by a logic as strict as any other application of sympathetic magic) it is necessary to consider the notions harboured by the Portuguese peasantry with regard to disease. These are almost as far removed from fact as is their belief in imaginary malign influences. Here again, the fault lies with that same fundamental inability to grasp the true relationship between cause and effect on which the whole edifice of magic is founded.

Nevertheless, it is impossible to dismiss as futile every folk-remedy that is not endorsed by Harley Street. If the root of the asphodel is used in the Serra de Monfurado as a specific against ringworm and other skin diseases, it is because the root contains free iodine and its effect is genuinely beneficial. In the same region it is usual, in a variety of diseases, to kill a young pigeon, cut it in half and clap the two halves while still warm and palpitating against the feet. In the particular case which came to my notice this treatment did not avail to save a child from death. Yet there must be cases in which it would have all the virtues of a hot poultice. Saliva, which figures in many folk-remedies, is known to have the properties of an enzyme and therefore to bring about fermentation. *Bafo de cão até com pão* ("Dog's saliva even on bread") is a popular saying. Inflamed eyes, for instance, are washed in water in which an ox has dribbled. A tube of mercury which is carried in many parts of Portugal as an amulet has been proved to be genuinely beneficial in cases of rheumatism.

The great bulk of folk-medicine, however, is founded on erroneous and superstitious notions. Illnesses are called by names which reveal complete ignorance of their nature and cause. For instance, a general weakness or decline, if not ascribed to *quebranto*, will as likely as not be diagnosed as *espinhela caida* (dropped sternum), and no pharmacopeia, however primitive, can admit that *o coxo*, the cause of eczema, can be cured by the hand that has slain a mole, or by the application by a fasting person who is careful to spit once on the ground and three times on the eruption, of oil containing the ashes of burnt straw or garlic. Nevertheless, hundreds of such remedies have been recorded. In the Alentejo, a person afflicted with warts is recommended to cross his hands and rub them against the ribs of a deceived husband without his knowing. In the Algarve, bloodshot eyes are cured by plucking onions or garlic without looking at them and then hiding them with the eyes shut. A headache is treated by placing on the forehead a copper coin covered with verdigris, and scrofula by scratching the affected part until it bleeds with a toe-nail from the left foot of a corpse and wiping away the blood with a cloth which is then placed in the coffin.

These and countless other remedies no less fantastic owe their origin to the fact that primitive man conceived illness not scientifically but animistically as some noxious beast, daemon or evil spirit which must be warded off, conquered or expelled.

I had actually penned these lines when my eye was caught by a paragraph in the *Diario de Noticias*, a Lisbon newspaper which showed that this conception of disease not only lies behind many folk-remedies which the unreasoning momentum of tradition alone keeps alive to-day, but is still actively held in parts of Portugal in the face of all scientific evidence to the contrary. The paragraph recounts that in September 1932 the little village of Carregueiras near Tomar was brought to the highest pitch of excitement by an old woman named Zefa, who claimed to be a *vidente* (seer) and to cure all human ills.

"The seer of Carregueiras", runs the account, "maintains that doctors fail through crass ignorance to cure their patients. They do not even know the nature of disease. Zefa knows and makes no secret of her knowledge. Disease, she states authoritatively (and Carregueiras in the main agrees with her), is a spirit. It is the spirit of some deceased member of the patient's family. When a patient consults her, Zefa does not enquire whether he has fever, headache, palpitations or a stomach-ache. She simply asks: 'Whom have you lost?' 'My father', replies the patient. 'Then that's what ails you', is Zefa's dogmatic pronouncement. And she adds: 'I will communicate with his spirit, and you will recover.' This she does, or so both Zefa and the patient maintain. And the patient is cured. At least, Zefa says so, and the patient says so too."

Padre Firmino Martins confirms the survival of this belief in his remote parish in Trás-os-Montes. "An unknown illness", he writes, "is a spirit or a soul from another world."

Since, therefore, all ailments arising from disease or from supposed evil influences are held to be the work of evil spirits conceived in a personal, almost concrete, form, the ignorant mind seeks to combat them with the same weapons as it would use against a personal enemy. With this object the peasant seeks to arm himself with powers as great as or greater than those possessed by his foe, and it is to the strength with which they are believed to imbue their wearer that a wide range of apparently disconnected objects owe their original selection as amulets.

Perhaps the most obvious application of this principle is the use as amulets of mirrors. For the glass which is powerful enough to reflect and repel (the primitive mind does not distinguish between these conceptions) so strong a power as that of light, can clearly ward off the *mau olhado* and the *mal de inveja* and turn them back on those who cast them.

It is a slightly different process of thought which regards the strength inherent in the amulet as being communicated by contact to its wearer, whose powers of resistance to disease it strengthens, like any modern tonic. One of the forces for which the primitive mind entertains the most unbounded respect is that of sex. It is scarcely surprising, therefore, that the perennial miracle of generation should be credited with unlimited powers, and it is these powers which the peasant seeks to harness for his own protection when he carries the "fig" symbolising the fecund union of the sexes, the "horn" an obvious representation of the phallus, and even, in some parts of Portugal, the actual genital organs of goats and other animals.

It is the celibacy and consequent sterility of the priesthood which account primarily for the superstition that priests are unlucky. A curious extension of this is the belief that lottery tickets should not be kept in oratories or near holy images.

The apotropaeic power in plants such as garlic and rosemary is doubtless derived from their powerful odour, and that credited to red objects is an extension of the belief in the virtue of fire and blood. A middle-class wife in Lisbon used regularly to burn rosemary in her house when her husband left for his work, with the avowed object of exorcising the evil atmosphere caused by their quarrels and bringing him back at night in a better frame of mind. The remedy cannot have proved efficacious, for she has now left him.

Other amulets probably derive their virtue from their value (silver objects, perforated coins) or rarity (four-leaved clover, prehistoric beads, or articles found in a peculiar concatenation of circumstances, of which instances have been quoted). There is doubtless a pre-Christian foundation beneath the use for the same purpose of objects deriving their virtue from consecration (holy bread or water, plants blessed by the priest). The power of the deity, whether Christian or pagan, is naturally regarded as being superior to the powers of evil.

The virtue of many amulets can be explained on the principle of the "hair of the dog that bit you". This very specific against hydrophobia is known in Portugal, and there are others based on the same principle, which is probably quite unconnected with the theory of homeopathic medicine. A viper's head is applied to snake bites, and it is believed that asthma is caused by cats and can only be cured by eating roast cat. Probably the fundamental conception underlying them is that expressed in the proverb: "Set a thief to catch a thief." If a knife is powerful enough to draw blood, how much more power must it have to staunch the flow of blood and thereby to protect its bearer from haemorrhage. A similar  line of thought accounts for the use of a "thunder-bolt" or of wood from a tree that has been struck by lightning as a lightning conductor, and of a half-moon to protect children from being moon-struck. In parts of Portugal, children are some-times given the moon as god-mother. It is better, the parents think, to enlist her on their children's side, rather than to risk her enmity. She is also invoked simultaneously in a single charm against toothache, fire and evil tongues. In the same way broom and the broom-stick, the very emblems of witchcraft, are used as a protection against witches, and it is probably a fundamental fear of the spirits of the dead which prompts the use as amulets of objects which have belonged to them in life.

Even more significant is the use as amulets against witches (as in the British Isles against fairies) of iron and steel. If witch-craft is, as has been supposed, the last flicker of altar fires dating from the remotest antiquity, it seems to follow that this super-stition is a dim memory of a time when the newly discovered metals were in concrete fact the decisive factor in a struggle with races whose only weapons were of stone or bronze. In Portugal, children are protected from witchcraft by a steel nail on the ground or a pair of scissors under their pillow. Open scissors or a steel knife are also effective in preventing the witches from overhearing any harm which may be said of them. The belief in the power of the modern weapon over the old faith is shown in the formula to be pronounced on meeting a witch:

Tu es ferro,	Thou art iron,
Eu sou aço,	I am steel,
Tu es o Diabo,	Thou art the Devil,
Eu te embaço.	I conjure thee.

By extension, steel objects are used as amulets against other evils which, in the first place, may have been regarded as witches' work. Steel rings are worn against gout, and a fragment of steel placed among a clutch of eggs prevents them from becoming addled. A remarkable taboo of the same nature was recorded in the Upper Minho in the middle of last century. The *Mordomo* (Master) of the Brotherhood of the Cross used to be forbidden, on the day of his installation, to touch any object of iron, on the score that this would bring down a pestilence on the village.

It is not, however, in concrete objects alone that virtue against baneful influences is thought to reside. A profound belief in the power of the spoken word accounts for the use of a wide variety of charms, spells, incantations, invocations and prayers, all of which are conceived as defending him who pronounces them against powers seen and unseen, and as themselves actively combating these powers. If, in a formula consecrated by tradition, the Devil is adjured to burst and all kinds of noxious vermin to disappear, the belief is that they will thereby be compelled to comply. Or the statement that they have done so will be thought to have the same effect. Such words may acquire additional power by being placed on the lips of Jesus Christ or a Saint, in the form of an anecdote or dialogue.

The power of the word is clearly evident in the manner in which prayers are often conceived. They are thought to achieve their object infallibly, by direct action like a spell rather than to depend on the pleasure of the deity or saint invoked:

Quem esta oração disser	He who pronounces this prayer
Não tema mal de maleitas.	Need fear no feverish ills.

There are charms, too, which accomplish their effect by imitative magic, such as the spell for ridding children of the *bicho* (insect) thought to be the cause of cutaneous diseases. The spell consists in saying: "Nine *bichos* on the child, eight *bichos* on the child, seven, ... six...." and so on down to "No *bichos* on the child", by which time it is confidently believed that the *bichos* will have been compelled to disappear.

There are incantations designed, by scaring away evil influences of all kinds, to secure the success of any undertaking, however trivial, such as setting a hen, erecting a scarecrow, or putting the dough to bake in the oven:

S. Levede te levede,	May St Leaven leaven thee,
S. Vicente te acrescente,	May St Vincent make thee rise,
S. João te faça pão,	May St John turn thee to bread,
Nosso Senhor te dê virtude,	May Our Lord endue thee with virtue,
E a nós a saude.	And give us good health.

Such spells, of which countless examples have been recorded, are often scarcely intelligible, much less translatable. They represent, for the most part, a crazy assemblage of formulas, derived originally from learned rather than from popular sources, such as medical recipes or mediaeval Latin prayers, which have become debased as they descended in the social scale, until in most cases the reciter himself would be hard put to it to explain them. The following is a characteristic charm against *ozipla* (erysipelas):

Pela Serra da Naia passei.	I passed by the mountains of Naia.
Bichos e bichas,	All kinds of vermin,
Sapos e cobras matei.	Toads and snakes I slew.
Santa Cecília encontrei,	I met St Cecilia,
Três filhas tinha:	She had three daughters:
Uma pela água abaixo,	One upstream,
Outra pela água acima,	Another downstream,
Outra foi visitar Nossa Senhora	The third went to visit Our Lady
E lhe perguntou que remédio lhe daria:	And asked her what remedy she would give:
Talha-l'a rosa vermelha,	Cut for her the scarlet rose,
Que le come e doi e proi	Let her eat it *e doi e proi*
Com sal do mar	With salt from the sea
E água da fonte	And water from the spring
E erva do monte,	And herbs from the hill,
Com o poder de Deus e da Virgem Maria	With the power of God and the Virgin Mary
E todos os santos e santas,	And all the Saints,
Em louvor de S. Pedro e S. Paulo,	In praise of St Peter and St Paul,
Em louvor de S. Silvestre,	In praise of St Silvester,
Que o que eu fizer tudo preste.	Let all that I do be of avail.

This incantation invokes not only the power of the Christian deity, the Virgin and the Saints, but also that of objects connected with pagan fertility cults such as salt, water and herbs. Others derive their power from the sun, moon or stars:

Estrela luzente	O shining star
Eu tenho uma ingua	I have a tumour
Ela diz que morras tu e viva ela.	Which says: Death to thee and life to it.
Eu digo que morra ela e vivas tu.	I say: Death to it and life to thee.

Many again owe a part of their virtue, if not the whole of it, to the introduction of magic numbers such as three (e.g. for the evil eye):

Se dois t'o deu	If two (eyes) gave it to you
Três t'o tiraram.	Three (persons of the Trinity) have taken it away.

seven, nine, or (in the prayer of the Guardian Angel), used against the Devil and witchcraft, thirteen, that is to say, the very number of witches in a "coven". Most charms to be effective must be pronounced three or even nine times. Thus, one of the many ways of getting rid of the *ar ruim*, is to go on three successive nights to a bridge, take water in your cupped hands, and, looking up to the sky, throw the water three times over your head, saying:

Ar e céu e estrelas vejo,	Air and sky and stars I see,
Se eu trago algum ar	If there is any "air" within me
Co'as minhas mãos o despejo.	With my hands I shed it.

In this last example one may see the clear intention of using sympathetic magic with the aid of the charm, to throw off the *ar* together with the water. A similar intention is to be discerned in most of the gestures which traditionally accompany the pronunciation of charms. Often indeed the charm itself consists of no more than a description of the accompanying action. The latter achieves its effect by sympathetic magic, and the power of the word makes assurance doubly sure. Thus impetigo round children's mouths is cured by the application of saliva and ashes, while the old wife recites:

Bicho, bichoucro,	Vermin, great and small,
Que vens cá buscar?	What seek ye here?
Con seiva da boca	With spittle from mouth
E cinza do lar	And ashes from hearth
T'o hei de queimar.	I will burn you up.

Then, mixing what has been in the mouth with what has been in the fire, she is sure, by contagious magic, of burning the disease in its animal personification.

The means considered up to this point of evading baneful influences, whether in the definite forms of disease, ill-fortune or a mysterious wasting away, or whether conceived merely as constituting an ever-present menace to health and prosperity, have been aimed at actively repelling them. But in so personal a form are these influences pictured that they are thought to be amenable to other and subtler forms of persuasion.

Aided perhaps by an incomplete understanding of infection, this personal conception of disease lies at the root of the belief that illnesses may be transferred to another person or animal. It is in this way that the cure for scrofula quoted on page 62 may be explained. The disease is clearly diverted from the living patient to the corpse and accompanies the latter to the grave. The same notion accounts for the custom by which a sick man carries on his person a lizard alive or dead, or a live spider immured in a nut. As the creature dies and withers, so is the patient supposed to be rid of his affliction. An Alentejan prescription for fever bids the sufferer mix nail parings with a little tobacco into a cigarette, which he must drop at a crossroads without looking up. The person who picks up and smokes the cigarette will then acquire the fever with it.

During a Portuguese summer, sunstroke is a not uncommon misfortune, and a number of other complaints are probably diagnosed as this when, in reality, they arise from other causes. The sun is thought to enter into the victim's head and to require removal therefrom. To accomplish this, a cloth is dipped in a glass of water and applied to the patient's forehead. As soon as the appropriate charm has been pronounced, it is said that the water begins to boil, for the sun has been transplanted into it.

So deeply ingrained is the conviction that evil may be avoided by diverting it to another person or object, that the recitation of folk-prayers is often accompanied by a string of curses on other people, and a charm for baking bread expresses the hope that the reciter's neighbour may have naught but horn to eat.

The spirits of disease are regarded as being equally susceptible to menaces, flattery or guile. To cure pins and needles the foot is threatened with St Thomas:

Desadormece pé,	Wake from sleep, o foot,
Que aí vem S. Tomé	For here comes St Thomas
Com um feixe de tojos	With a bundle of gorse
P'ra te queimar os olhos.	To burn your eyes.

Fevers are addressed as "friends" and bribed to go away:

Amigas, ide-vos embora,	Fevers go hence,
Levais pão para comer,	Take bread to eat and
Palha para vos deitar,	Straw for your bed,
Adeus que vos não quero tornar a ver.	Farewell and never let me see you more.

In another case not only bread and straw, but wine to drink and a basin to wash in are left beside a tree, and the patient having pronounced the charm, flies without looking back, convinced that the fevers will be satisfied to remain where they have all a man can need, unless of course some other person takes these offerings, when the fevers will accompany them.

The prettiest example of guile is the method used in the Algarve for curing a stye. "In the King's name", cries the patient, "the house of my stye is on fire." And to trick the stye into hastening away to the scene of the conflagration, dried grasses are burnt and the affected eye is held over them. Another trick is to pretend that a remedy has been efficacious on a previous occasion. To cure oneself of sticky hands, all that is necessary is to enter a church which one has never visited before and say:

A primeira vez que nesta igreja entrei	The first time that I entered this church
O suor das minhas mãos aqui deixei.	I left here the sweat of my hands.

Such innocent things as *herpes* (tetters) may be driven away by saying to them, on an empty stomach: "As sure as I have just eaten and drunk, been to Rome and come back again, so surely may you remain with me."

These benighted beliefs and practices owe their survival very largely to the prestige enjoyed by a prelacy of men and women of virtue who not only preach the tenets of magic but also derive considerable profit from their practice. This hierarchy includes a wide variety of witch doctors, variously named *bruxos, feiticeiros, mestres curandeiros, saudadores, bentos, fadas, benzedeiras, mezinheiras* and *responsedeiras.* These names are apt to carry slightly different meanings in different districts. The name *bruxa* may denote a wise woman rather than a witch, and *feiticeiro* or *feiticeira* may as easily mean a person who uses his magical powers  for good as for evil. An Alentejan woman made a distinction between these two terms which is enlightening, even if it is not universally accepted. "One is a *bruxa* by nature", she said, "and a *feiticeira* by arts, while it is fate which makes one a *lobisomem* (werewolf)." The Algarve fishermen are accompanied to sea by a *bruxa* who tells them when and where to let down their nets, and who is rewarded with an equal share of the fish.

"In spite of all persecutions", wrote a contributor to *Trás-os-Montes*, a local organ of that province, "the *feiticeiros* continue to ply their trade, principally in the villages, but also to a lesser degree in the towns. A man falls ill, and his relations have recourse to the *bruxo*. They send for the latter or lift the invalid, often at the point of death, on to a mule, and set off over the mountains along precipitous paths which would be difficult even for the hale and hearty. They take with them the customary presents of wine and oil, chickens and cheeses, which they humbly offer to the *bruxo* or *fada*. These, with a grave air, are so gracious as to comply with the request made to them, and prescribe the most ridiculous remedies to cure the gravest illnesses."

"In a village near Sardão, a boy has been ill for a year with a tumour on his leg. When his relations had applied the usual *mezinhas*, all to no effect, they took him to the Doctor who recommended him to go to Oporto for an urgent operation, failing which the leg would probably have to be amputated. This advice did not prove to their liking. A journey to Oporto would cost money, and the *feiticeiro* was just as good. So they went to the nearest wizard who pronounced the following oracle:

"'The boy can be cured, but you must carry out exactly all that *I* tell you. You must kill a lizard. But note that it must be killed with the index finger with slow and regular blows. When it dies, place it in an egg-shell, and make the boy piss into the shell until it is full. Then leave it for three days in the smoke of the hearth, after which you must anoint the injury with the liquid using the lizard as a brush.' These prescriptions were carried out to the letter, and by chance the tumour temporarily healed. General wonder and astonishment: praise and many gifts to the *feiticeiro*. Eight days later the wound opened anew, and it is probable that the leg will have to be amputated."

Additional impetus was given to the war against witch doctors by an incident which sent a thrill of horror through the country in February 1933 and was fully reported in the newspapers. In the hamlet of Oliveira near Marco de Canavezes some forty miles up the Douro from Oporto, an epileptic woman was thrashed into insensibility and then burnt to death in order to exorcise the evil spirit which was thought to possess her and which was stated by a *bruxa* to have caused the illness of a neighbour. This ghastly *auto da fé* was accompanied by the reading of exorcisms from the Book of St Cyprian,[1] and the peasants who conducted it were convinced that the victim would arise Phoenix-like from the flames, liberated of the spirit which had troubled her.

It can readily be imagined that warfare against these witch doctors is waged not only by the Church and the medical faculty, on whose preserves they trespass, but also by the authorities responsible for the welfare of the peasantry. Not so long ago forty *bruxas* were arrested in one day at Gaia.

Whatever the country may lose in picturesqueness through the disappearance of the *bruxas* will be as nothing to what it will gain in improved health and living conditions. No longer

---

[1] St Cyprian is popularly supposed to have sold his soul to the Devil, by whom he subsequently became possessed, and to have been exorcised by a holy woman whom he married. His book, a farrago of nonsensical and pernicious incantations and lore, is widely distributed in the north in a recent edition. It is used among others by treasure seekers who describe a circle or pentacle where *coisas de mouros* (antique remains) are found, and, standing within it, read charms from the book.

will witch doctors order a desperately ill woman to be laid on the ground and have cattle driven over her (an authentic case which failed to save the patient if it did not actually contribute to her death); or a priest to throw a bone three times over a church tower in order to rid himself of the persecution of witches. No longer will dead men's bones dissolved in wine, or the water from seven different springs be prescribed for fever, when quinine is all that is needed.

The episcopal *Constituições* of Braga (1639) prohibit the use of set formulae to influence the course of events *ainda que seja por meyo de orações...feitas a Deos nosso Senhor* ("even though it be by means of prayers to God our Master"). Similar proscriptions are repeated in the *Constituições* of Oporto (1687) which sentence offenders to be placed at the door of the cathedral with a *corocha* (sort of dunce's hat) on their head on a Sunday which is also a Patron Saint's day, and thereafter to be exiled. This penalty is imposed for such offences as "forming fantastic apparitions, transmutations of bodies, voices that can be heard without being seen, and other things which exceed the efficiency of natural things". In truth, the twentieth century would have received short shrift at the hands of the seventeenth.

It is, indeed, only thanks to the credulity of the peasants that the witch doctors have survived centuries of persecution both ecclesiastical and civil. To that credulity there would seem to be no limit.

At Caneças, on the very outskirts of Lisbon, a doctor was called in to see an old woman, and began by asking her where she was in pain. *É o Senhor que sabe,* was the unexpected reply. It was his business to know, she said, not hers to tell him, and she sent him about his business. Clearly her previous medical advisers had affected universal knowledge.

At Tondela (Beira Alta) a bride disappeared on the eve of her wedding. The groom went with the girl's relations to consult a well-known witch at Nelas. On entering her presence, and in order to test her powers, they first enquired, with peasant cunning, if she could tell them on what errand they had come. "If all my clients were so exacting", was the witch's answer, "I should do little business." Evasive as it was, this reply was nevertheless sufficient to lull their suspicions.

At Entradas, early in 1931, according to a newspaper cutting in my possession, a young girl whom her parents noticed to be growing daily thinner and more wan, was sent to consult a certain Salvador Antonio, an old man of eighty-one who enjoyed great fame as an exorciser. The witches had entered into Maria Antonia, said the *feiticeiro*, but he would soon expel them from her. With this object he pronounced incantations (*benzeduras*) over her on Tuesdays and Fridays, and at the new moon. When this treatment had been followed for some time, Maria Antonia gave birth to a child. Clearly, this was the form in which the witches had elected to depart from her person.

Nevertheless, not all these witch doctors are conscious charlatans. In particular, *bentos* and *bentas*, who derive their reputed powers from the fact that they have cried aloud in their mother's womb,[1] are often quite convinced of the reality of their gifts. Leite de Vasconcelos has left a memorable account of one such man, which may be quoted in full:

All the people in the neighbourhood as far as Lamego called him in for their illnesses. He had a grave manner and a slow, gruff voice, like a prophet's. But he was very fond of *licor de São Martinho*. When he was summoned, he would mount his ass, arrange the saddle-bags in front of him, sling a *santo-christo* round his neck and then set out to cure suffering humanity. His prescriptions were in no way different from those of charlatans: infusions of dried herbs, decoctions prepared from the burnt shirts of his patients, a few prayers and no more. On various occasions the law had interfered with his sacred functions. But neither the black looks of the Judge nor the sombre walls of the prison-cell had been able to wean him from his mode of life. For he had wept in his mother's womb. Everywhere he received evidences of his "virtue". From afar men held out their arms to him. At home there was always a crowd of patients round his door, as I have seen more than once. What more could he need? He never accepted money, but he would not refuse fruit, meat and so on. It was for this that he always carried saddle-bags slung across his ass. Sometimes the priests of various parishes attacked him. But he, ever firm in his predestined mission, would never answer more than: "I am a *bento*, and you are not."[2]

---

[1] One Luis de la Penha, charged with sorcery at Évora in 1626, confessed that he "understood and divined things because he had cried in his mother's womb"

[2] Leite de Vasconcelos, *Ensaios Ethnographicos*, II, p. 157.

It is not only for their powers of healing or of exorcising evil spirits, that the Portuguese wizards are sought out, but also for their powers of foretelling the future and of influencing it.

The belief in divination is as firmly rooted in Portugal as is the belief in charms. In Lisbon the oracle may be consulted in the following manner: take three cocaine (*sic*. poppy?) seeds in your hand, cover your head and step out on to a balcony. As the clocks strike the hour, say:

> Three corpses out of three graves,
> Three souls will join them,
> Three drops of blood also,
> In the voice of the people let me hear my answer.

The first actual scrap of conversation overheard from some passer-by must then be interpreted as the answer to your question.

A girl may learn if a young man loves her by putting a stone in his shoe. If, as appears probable, he removes it, he is unmoved by her charms. But if the stone is small enough or the swain patient enough to suffer its presence, there is nothing he will not suffer for her sweet sake. Some auguries of this type partake of the nature of spells in that their use is believed not only to procure the answer to a question but to ensure that this answer is favourable. A strange charm from Cumieira near Regoa, pronounced at midnight for the purpose of winning a husband, begins by invoking St Helena,  expresses the wish in question and concludes: " If it shall be so, let my dreams be of clear waters, newly washed linen and green fields: if not, then let them be of dry pastures, dirty linen and turbid water."

No less than charms such auguries have fallen within the ban of Church and State.

The *Constituições* of Braga (with special reference to gypsies), Oporto and Guarda condemn among other pagan practices "auguries, making conjectures by voices, by meeting animals, by the singing or flight of birds or similar signs". Nevertheless, the Portuguese peasants persist in seeking to know the future and to influence it according to their desires, and to this end they invoke the assistance of witch doctors with their spells and potions. A story from the Azores tells of a girl who sought to win a young man's affections by kneading dough against her bare breast, baking it and offering it to him. The man had his suspicions and gave the loaf to his donkey, with the result that the poor lovelorn beast never quitted the maiden's door. Two love spells collected by Leite de Vasconcelos in the North are worth quoting. The first, learnt from a witch at Guimarães, is intended, for all its savage phrasing, to attract the tender looks of a young man:

Com estes dois te vejo,	With these two (eyes) I see thee,
Com estes cinco te arremato,	With these five (fingers) I bind thee,
O coração te trinco	I split thy heart
E o corpo te parto.	And I cleave thy body in twain.

The other is intended to extract a declaration of love and is accompanied by action suited to the words:

Assim como eu pico este limão,	As I prick this lemon,
Assim pico o teu coração,	So do I pierce thy heart,
Que não possas comer,	That thou mayst neither eat,
Nem beber,	Nor drink,
Nem dormir,	Nor sleep,
Nem descansar,	Nor rest
Enquanto não vieres falar.	Until thou hast been brought to speak.

So, also, the witches teach that a young man, to win a girl's love, must pass a threaded needle first through a snake's eyes and then through the girl's clothes, or touch her with the skeleton of a serpent which has been left in running water till the flesh has rotted away.

Of the magical beliefs and practices which have been described in this chapter, many are blended with elements derived from Christianity or from the anthropomorphical religions which preceded it. As has been shown, the belief in magic to the exclusion of a supernatural divinity, can only have existed, if ever, in the infinitely remote past of the human race. Nevertheless, it is possible to trace in Portuguese folk-belief the gradual evolution of religious thought through animism and polytheism to anthropomorphic monotheism.

Of the first stage, involving the conception of trees, rivers, rocks and mountains as being imbued with a life of their own, there survive, as may be expected, only a few examples. A lady who may read this book will, for instance, recall how an old woman in the Serra de Caramulo greeted her with the wish *que a benção da montanha os acompanhe até casa* ("that the blessing of the Mountain may accompany you home"). In the district of Chaves, near a mountain rising out of the Barroso uplands which still bears the name of Larouco, a votive tablet has been found with a Roman inscription rendering thanks to the god Laroco for the recovery of a sick husband. The personal conception of a river is evident in the Minho adage that *o rio Ave há de comer um fôlego vivo por dia* ("the River Ave must devour a human life each day").

Far commoner throughout the country is the belief that rocks, springs and mountain caves are the dwelling places of beings in human form who correspond to the nymphs, sprites, nereids and other nature divinities of antiquity. Here is polytheism in its purest form. Mothers will tell their children that at the bottom of the well is an old woman waiting to devour them. Generally, however, these beings are given the name of *mouras encantadas* (enchanted Moorish princesses). They are associated not only with natural phenomena such as may have lain behind primitive religions, but also with material vestiges of ancient races. They haunt ruined cities and castles, and old worked-out mines. To the peasants archaeological remains are *oisas dos mouros* ("things of the Moors"). Dolmens, of which there are several fine examples in the country, notably near Guarda and Montemor o Novo, are called *lapas* (caves) or *pedras dos mouros*. There is a folk-tradition that the wonderfully carved *Pedra Formosa* from Citânia de Briteiros, the pre-Roman settlement near Guimarães, was carried by a *moura*, spinning as she went, from S. Romão to·S. Estevão.

At the highest point of the Serra de Monfurado, which in the low-lying Alentejo is a landmark for miles round, stands

a boundary-stone, some ten feet high, apparently of great age, with small square cavities hewn in one side. The stone has snapped in two, about two feet from the base, and the longer part now lies horizontal on the ground. This stone, say the peasants in the surrounding cork forests, is the tomb of an enchanted Moorish princess.

Mouras encantadas are conceived as beautiful maidens, having sometimes a snake's tail in place of lower limbs. To mortal eyes they are visible only on St John's night, when they may be seen combing their hair like mermaids with golden combs, spinning or weaving with golden thread, or laying out figs in the dew, which turn to gold in the hands of any mortal who can seize them. They are almost invariably the guardians of a treasure left by the Moors which may take the form of a stall full of rich jewels or of a buried hoard of gold.

They are not maleficent, and their contacts with human beings seldom end to the disadvantage of the latter. They are shown rather as requiring human aid to break their enchantment. In the many legends in which this aid is invoked they are often portrayed in serpent form. Near Vernoim in the Minho, runs one such tale, there was once a serpent which struck fear into the hearts of all around. A young man who went forth to slay it fell asleep during his quest. The serpent kissed him while thus he slept, and was transformed into beautiful maiden whom he married. More usually it is the mortal who must kiss the serpent and is at the last moment overcome with an invincible repulsion, thus doubling the spell instead of lifting it: or who, being asked to decide between the serpent's eyes and its treasure, misguidedly chooses the latter with the same unfortunate result.

When the *moura encantada* is in human form the means of disenchanting her are simple. But the omission of some trivial detail usually renders them vain. Near Turquel a *moura* asked a herd girl to bring her a cake baked without salt in a new oven from dough kneaded in a new basin with fresh spring water. The cake which the girl brought fulfilled none of these conditions, and as the *moura* ate it, a sad look came into her eyes and she disappeared with the traditional cry of *Ai, que me dobraste o meu encanto* ("Alas, you have doubled my spell"). Another *moura* once gave a man three small loaves, bidding him place them on a certain rock and walk round it thrice. The cakes would then be transformed into three steeds, one for himself, one for the *moura* and the third for her treasure. Unfortunately, the man's wife found the cakes in his saddle-bag and ate them, thus bringing his plans to naught. Equally unsuccessful in removing a *moura's* spell were a girl who threw a hot cake into a fountain a few minutes later than the appointed hour of midnight, and another who, bidden to bake a little horse of dough, had the misfortune to snap off one of its legs.

The *mouras encantadas* have points in common not only with the Nereids of antiquity but also with the Celtic fairies and Basque *lamiñak*. Like these they dwell in barrows, and are credited with the construction of great buildings in a single night, such as the cathedral of Viseu and the Romanesque church of Leça do Balio. With them, also, they share, in a widely distributed legend, the services of a human midwife, and the familiar confusion in their gifts between gold and dross.

A midwife from near Alcobaça was once summoned into the country and conducted through a narrow and scarcely perceptible opening into a cave. In reward for her ministrations she was given a small piece of tile. Some time afterwards a passing beggar showed her that the tile had turned to gold. A chain of lumps of coal given by a *moura* to  a little girl was similarly transmuted into precious stones. More usually, it is the other way round, and presents of gold are found to have changed into coal.

The identification of *mouras* with serpents suggests analogies of another kind. It is believed in South Africa, India and was also in Ancient Greece, that the souls of the dead appear in serpent form. The treasure buried with the dead in ancient times, the white sheets worn by *mouras encantadas* in certain districts, and the archaic jars said to have been found where they are reputed to exist, are all links in a chain leading from these unhappy princesses to the spirits of the dead.

In this respect, as in many others, the Christian Church has built on pagan foundations. Though many *mouras* remain, as many more have been displaced from their rocks and fountains by chapels dedicated to the Virgin. In such cases Our Lady sometimes inherits the attributes of her pagan predecessors. Near Arcos de Valdevez there is a rock on which the Virgin is said to sit spinning; and the same idea is reflected in a couple of popular quatrains:

A Senhora da Lapinha	Our Lady of the Little Cave
Anda no monte sem roca,	Wanders on the mountain without a distaff,
Para acabar uma meada	To finish a skein
Falta-lhe uma maçaroca.	She lacks a spindle-full.
A Senhora da Lapinha	Our Lady of the Little Cave
Tem uma meada de ouro	Has a golden skein
Lavada na Fonte Santa	Washed at the Holy Spring
Côrada no Miradouro.	And dyed at the Miradouro.

The association with the Moors, in other countries as well as Portugal, of vestiges of ancient races, rites and religions is a phenomenon of which other instances will be given in a later chapter of this book. It is deserving of specialised research and study, since it raises problems difficult to solve. It may be suggested that, just as the modern Greeks have preserved a dim folk-memory of the ancient Hellenes and apply this name to-day to beings closely resembling the Portuguese *mouras*, so have the Portuguese preserved a distorted recollection of the last great conquering race which overran their land. There may perhaps be no more in it than this. Yet this solution does not seem to account satisfactorily for all the aspects of the problem, and I, personally, am confident that some other may be found. A more hopeful line of enquiry is opened up by a suggestion put forward by Martins Sarmento that the word *mouro* came to be a synonym of "pagan", since it is applied

Plate V

THE CASTLE OF GUIMARÃES

SAO MAMEDE'S DAY AT JANAS

Plate VI

in certain parts of the country to unbaptised children, and, by extension, to undiluted wine. In Basque folk-lore, *lamiñak*, "Moors" and "Gentiles" are inextricably mingled.

Portuguese folk-lore also contains reminiscences of various other supernatural beings; monsters such as *olharapos* (one-eyed cannibal giants like the Cyclops); *Almanzonas* (Amazons); giants and dwarfs; and ill-defined horrors such as the *Medo, Pesadelo, Trasgo, Tardo* and others.

This chapter would not be complete without some reference to another form of supernatural being which, although it is closely connected with the belief in witchcraft, cannot be completely identified with it. I allude to the belief in *lobisomens* (man-wolves), once widespread throughout Europe, and still very much alive in the north of Portugal. The Portuguese name for these creatures is the equivalent of the English "werewolf", for "were" is the Saxon *wer*, a man (although an ancient manuscript preserved in the Bodleian Library offers an ingenious alternative etymology: "Ther ben somme that eten chyldren and men, and eteth noon other flesh fro that tyme that thei be a-charmed with mannys flesh, for rather thei wolde be deed: and thei be cleped werewolfes for men should be war of hem").

The belief current in England three centuries ago is thus summarised in Richard Verstegan's *Restitution of Decayed Intelligence* (1605): "The wer-wolves are certain sorcerers, who having anointed their bodies with an ointment which they make with the instinct of the devil, and putting on a certain enchanted girdle, do not only unto the view of others seem as wolves, but to their own thinking have both the shape and nature of wolves, so long as they wear the said girdle; and they do dispose themselves as very wolves, in worrying and killing, and most of human creatures."

The Portuguese *lobisomem*, though the object of superstitious fear and abhorrence, is on the whole a more harmless creature. Unlike the witch or the sorcerer he does not become a werewolf for choice or in pursuit of malefic ends. The Portuguese belief, unanimous in this one respect, is confused and conflicting as to the manner in which werewolves are made.

In some parts of the country they are thought to be men who have died the victims of witchcraft or at whose baptism some of the responses were omitted. Elsewhere *lobisomens* are held to be the children of incestuous parents or the fruit of marriage between *compadres*, those linked by ties of godfathership.

The most widely held belief is that the seventh (or sometimes fifth) consecutive son or daughter is fated to turn into a werewolf, unless the eldest becomes his godfather or godmother or pricks blood from his little finger, or unless the child is baptised with some special name such as Adam, Eve, Jeronyma or Bento. Should these precautions be neglected the child will, usually about the age of thirteen, leave his home and go out into the woods. There he chooses a tall pine tree, often near a brook, at the top of which, out of all men's reach, he leaves his clothes. Then, dropping naked to the ground, he finds a spot where some animal has rolled and, lying there, will assume the shape of that animal, whether wolf, donkey, dog, cat or even hen. From that day he begins to "run out his spell" (*correr o fado*). On certain nights of the week, usually Tuesdays and Fridays, but sometimes Wednesdays and Thursdays, he assumes animal form, and gallops like the wind across the countryside, taking fields and hedges in his stride. So great a noise does he make that he can be heard on the stormiest of nights. His first night's course must take him by *sete montes, sete pontes e sete fontes* ("seven hills, seven bridges and seven springs"). Thereafter he is obliged on each occasion to pass through seven parishes or seven fortified cities (*sete vilas acasteladas*). Then returning to the pine tree, he resumes his clothes, his human form and his normal life. The day finds him exhausted, but inconscient of his nocturnal escapades. From other men he is to be distinguished only by his leanness, his yellowish tint, and by blisters on the hands that have served as paws the whole night through.

The enchantment may last only for seven years, but usually, unless it is broken at an early age, it is lifelong. There is only one way to break the spell (*quebrar o fado*). The blood of the *lobisomem* must be drawn, preferably with steel. Thus, at Vila Cova, Leite de Vasconcelos was told the name of a man who had slashed a werewolf son, and thus freed him from his *fado*. In some places the wound is thought to be effective only on the tail, which corresponds with the little finger of the victim who, when he resumes human shape, is distinguished by the lack of this member. The greatest care must always be taken to avoid being touched by the spurting blood, for anyone so touched will himself become a werewolf. The only person who can be touched with impunity by the blood of a *lobisomem* is his wife. To all other persons his blood, like his bite or even his saliva, is fatal.

Literary references to werewolves are first found in Portugal in the sixteenth century, but the belief in lycanthropy, whether indigenous or not in Portugal, goes back to the classical age in other countries. Hermann Urtell relates it, in both its general and its Portuguese forms, with a vegetation myth; the leaving of the clothes in the pine tree near water followed by an alteration in the ordinary course of nature linking it, in his opinion, with the cult of the spirit of vegetation. This interpretation is scarcely convincing, yet it is difficult to imagine any more plausible. Many details of the Portuguese belief, such as the use of steel, the werewolf's fear of cockcrow, his preference for certain nights, seem to connect it with the witch cult. But this may be no more than the usual confusion in peasant minds, linking in a common anathema all vestiges of pre-Christian practice and belief. It is possible, nevertheless, that the adoption of animal disguises for ritual purposes, and the acting of animal parts which were not always clearly distinguished from actual transformations, lie at the root of the belief in lycanthropy. This, at least, seems more acceptable than the theory which sees in it "a form of hysteria and a pathological condition (frequently recorded in pregnant women) manifesting a depraved appetite and an irresistible desire for raw flesh, often that of human beings, frequently accompanied by a belief on the patient's part that he or she is transformed into an animal". As well might one regard as a case of lycanthropy the child at a certain unconventional school who recently developed a wolf complex and was allowed to live out in the woods for a fortnight.

CHAPTER IV

BIRTH, MARRIAGE AND DEATH

Just as, in Portugal, people are held to be more subject to malign influences at the "open hours" of sunrise, noon and sunset, so, in the same way, perils to soul and body tend to accumulate about the critical moments of birth, marriage and death. It is hardly surprising, therefore, that these milestones on the path of life are attended by innumerable superstitious beliefs and practices.

Even in civilised societies, where the thought and observation of countless generations have at long last succeeded in relating cause to effect, the dangers attendant upon these occasions are fully realised. Birth is fraught with peril to mother and child. Marriage is not to be entered upon lightly. Death is the threshold of an unknown and awe-inspiring world. Round these three events centre the major tragedies of human life. If, in our enlightenment, we still approach them with care and caution, how much more will the peasant, often ignorant of the true cause of his disasters, seek, on these occasions, to conjure the perils which beset his path.

No condition is more perilous than that of a woman with child. In parts of Portugal there still survive memories of the taboos with which among primitive races she is surrounded, and of which the majority may be explained by the principles of sympathetic magic. At Arcozelo a pregnant woman must not sniff flowers, for if she does so her baby will be disfigured with birthmarks, called *flores* in Portuguese. If she passes beneath a wire or the cord by which some animal is attached, the infant will be born with the umbilical cord wound round its neck.

To resolve the absorbing problem whether the baby will be a girl or a boy, several methods of divination are practised, all picturesque rather than scientific. One such method consists in spitting on a chestnut and throwing it into the fire. If it bursts the child will be a boy; if it only fizzles feebly it will be a girl. An alternative is to note with which foot the expectant mother descends the first step of a flight of stairs; if with the right, a boy will be born; if with the left, a girl.

To ensure an easy delivery a woman may walk under the canopy in a religious procession. When her hour draws near, nine virgins, all named Mary, should ring nine peals on the church bell, holding the rope with their teeth. To hasten child-birth, a good method is to cut a thread of twisted red silk into fragments, and drink these in wine. When the pains start, in the Alentejo a man's hat is placed on the mother's head until the child is born. The anxious husband can be of more use than is usually the case; for he can hasten a slow or difficult delivery by lifting and turning over a tile from a church roof.

Of such pre-natal observances, the most striking is the *baptizado à meia noite*, still practised on rare occasions in the Minho. Should an expectant mother for any reason fear a difficult delivery, or should a previous child have been stillborn (to quote an instance from Ponte da Barca), she may go to the middle of a bridge at dead of night accompanied by a male relative other than her husband and draw a bucket of water from the river. They then await the first man to cross the bridge after midnight and oblige him, if necessary on pain of death, to baptise the unborn child. The woman lowers her garments, and the stranger, dipping his fingers in the water, makes the sign of the cross on her bared body. Incredible as it may seem, I have it on unimpeachable authority that there lives in Barcelos to-day a person who was baptised in this manner. At a village in the Gerez Mountains it is the priest himself who performs the baptism in the bed of a rocky stream.

This rite clearly derives its virtue partly from the good fortune traditionally associated with a "first-footer" and partly from the conception of water as a fertilising as well as a cleansing agent,[1] a conception reflected in the tradition round Lisbon that a bride's back should be washed with water on her wedding-night by one of her relations "for luck"; and in the widespread prejudice in favour of rain falling on a wedding-day.

There is no end to the precautions which must be taken to protect a newborn child from the perils which beset him before baptism. The baby is called a *custódio* (? guardian angel) or, as mentioned in the previous chapter, a "Moor" until he has been baptised into the Christian faith. If he dies unchristened he goes "into the limbo". During this period a lamp must always burn in the room, and the child should not be swaddled in a sack lest he grow up into a thief.

Special precautions are taken on a child's first contact with the element of water. At Guimarães a cross is made over the first water in which he is washed. Needles are thrown into the water and a special benediction pronounced:

Aguinha a lavar	Water for washing
O Senhor a benzer:	And the Lord will bless:
Àguinha a correr	Water running
E o menino a crescer.	And the child will grow.

It is also thought good to throw a coin into the water in order that the child may grow up laborious. After use the water must be thrown away; that of a boy outside the house, and that of a girl indoors, for a woman finds happiness in the home while a man seeks it abroad.

While still a *mouro* a child is thought to run special risk of being carried off by witches. There are several ways of protecting him. One, observed in parts of the Algarve, is to give him the pre-baptismal name of Inácio or Inácia. Garlic is sewn

[1] In Barroso, a childless couple will repair to the lonely bridge of Misarela and standing in the bed of the river await the first man to cross the bridge after midnight. Him they will invite to be godfather to the child which they are now convinced will be born to them.

into the baby's clothes with the same object, and a still more curious charm is to slip the father's trousers over the infant's arm. His first shift is kept unwashed, and if witchcraft is suspected, is set to boil on the fire, whereupon the witch will come in through the door and endeavour to remove it. Rue and rosemary fastened to the bed will also protect the child from being carried off by witches.

Even after baptism there are still many taboos to be observed. If the cradle is rocked in the absence of the baby, the latter will grow up wild and lawless. Such a child can be cured in one of two ways. He can be taken to S. Marcos da Serra in the Algarve and have his head knocked against that of the bull lying beneath the feet of the Saint's image, while the mother recites:

Meu Senhor São Marcos	My Lord St Mark
Que amansas touros bravos,	Who tamest wild bulls,
Amansai-me este filho	Tame me this son
Que é pior que todos os diabos.	Who is worse than all the devils.

Or he can be laid upon the altar of a chapel dedicated to S. Gonçalo at Penha d'Aguia in Trás-os-Montes, while the godmother walks nine times round the chapel, after which she takes the Saint's crozier and applies it nine times to the child's posterior.

The fear of dumbness is peculiarly widespread. It is believed that a child will never talk if, before he has begun to do so, he sees his reflection in a mirror or has his hair cut. To make a dumb or backward child speak, his godmother will carry him in a sack on her back to three (or seven or nine) houses. At each of these she will beg on his behalf:

Esmolinha para a creança do fole	Alms for the child of the bag[1]
Que quer falar e não pode.	Which would speak but cannot.

She will eat a little of whatever is given to her and give the rest to the child, taking care to leave by a different door from that

[1] The word *fole* means both "bag" and "lung", and there is a clear connection between the bag and the child's lungs.

by which she entered. Another and doubtless equally effective method is for the child to pass, with a bag of sweets in his hand, beneath the platform on which the statue of S. Luís is carried in a certain procession at Braga.

A child's growth is particularly liable to be affected adversely by "sympathetic" influences, depending for their effect on the principles of contagious magic. If his clothes are wrung out in the wash, his growth will be twisted. If another child jumps over him in play he will never grow any taller.

For the nervous child a treatment is prescribed which is more picturesque and not necessarily less effective than that advocated by psycho-analysts and disciples of Montessori. Such a child can be cured either by taking a black cock in offering to São Bartolomeu do Mar near Esposende, or by making him hide behind a door and eat a roasted cock's comb.

In view of the sympathetic sensitiveness of all severed portions of the body, it is natural that care is taken of the child's first teeth when they fall out. The surest method of averting any disaster to them and therefore to their owner is to throw them over one's shoulder or on to the roof or oven, repeating three times some such formula as:

Dente fora	Tooth away
Outro melhor na cova;	A better one in its place;
Em louvor de S. João	In praise of St John
Que me dê outro melhor	May he give me another better one
Para comer o pão.	Wherewith to eat bread.

If all these precepts and taboos are religiously observed it is just possible that the infant may survive the many perils of child-hood and attain the scarcely less "dangerous age" of adolescence. The time will come when he or she is ripe for marriage.

Marriage in peasant Portugal is neither purely a matter of sentiment, nor yet one of convenience alone. If marriages are arranged by parents rather than by those most directly concerned, affection plays a considerable part in the choice, and the innocent delights of courtship are not frowned upon. In Portuguese, courtship is called *namoro* or *derrete*, this latter word being derived with picturesque aptness from the verb *derretir* (to melt). Hence the famous *muro do derrete*, the "wall of melting glances", now sacrificed on the altar of progress, which was once the principal feature of the October fair at Mercês near Cintra. Along this wall, on a row of embrasured seats, all the marriageable girls of the surrounding countryside were ranged with their dowries in the form of golden coins hung round their necks. The eligible young men used to inspect the goods thus set out for sale, guided no doubt by the principle that "handsome is as handsome......has".

A courting couple are known as *conversados*, and people tactfully leave them to themselves, within the bounds permitted in a southern clime. They exchange the gifts consecrated by tradition: a ring, a rosary, a handkerchief embroidered with a heart, a name or a tender verse:

Vai-te lenço venturoso	Go, lucky handkerchief
A um *sincivel* coração,	To a sensitive heart,
Vai levar afectos meus	Go, carry my affections
A quem é minha prisão.	To her who holds me captive.

But, since handkerchiefs may be used to dry the tears engendered by absence, such ill-fortune must be averted by accompanying this particular gift with the present of a coin. In the same way a Saint's image or a rosary must not be directly offered by one lover to another. The recipient must merely be told where they lie, and must find them for himself, thus preserving the pretty fiction that they have not been given but stolen.

Within recent times the wedding ceremony has lost many of the more picturesque and curious observances which tradition, in this as in all other lands, has accumulated around it. The following account, from the pen of João Eloi, of a wedding in the Pombal district, illustrates many of these attractive features:

Having received the paternal benediction, the bridegroom, the best man and the other guests went to the bride's house, the groom wearing his cloak, a custom which is so deep-rooted that however hot the day the groom never fails to wear this cloak.

The door of the bride's house was closed, and there was no sign of life within. One of the *padrinhos* (godfathers) stepped forward and knocked on the door, the following conversation ensuing between him and the bride's mother:

"Who goes there?"

"People of peace and concord."

"What do you want?"

"We have come for three things: first we have come by the grace of God and of the Holy Spirit; secondly for our honour and your profit; and thirdly to take away Teresa Gonçalves who has betrothed herself to our godson Custodio Alves, that she may go in our company to enter the holy order of matrimony."

"You have mistaken the door. Try higher up or lower down."

The groom then intervened:

"I know very well where I am, for I have come many times by day and by night and I have never mistaken the door."

Then one of those within opened:

"Since the bride's mother permits, you may enter and take her."

Then entering, followed by all, the groom said to the lady of the house:

Nesta casa vou entrando,	I enter this house,
Não me chamem atrevido,	Do not call me over-daring,
Venho buscar uma pessoa	I come to seek a person
Que me tinham prometido.	Who has been promised to me.

But the bride was not there, and he was told that he must look for her and that if he could find her she should be his. With two wedding cakes in his hands, the groom began to look for the bride whom he finally found in a dark room with her godmothers and father. Giving a cake to each of the latter he cried:

Deus as salve a todas três,	May God save you all three,
A todas três que aqui estão.	All three who are here.
A todas trago na alma	You are all in my soul
Só uma no coração.	But only one in my heart.
Menina que há de ser minha	Maiden that art to be mine
Dê-me cá a sua mão.	Give me here thy hand.

Then holding hands they went to the bride's mother at whose feet they knelt down, both asking her pardon, the groom adding: "If hitherto I have called you *Tia Maria*, I promise henceforward to call you Mother."

The pardon and the blessing having been given, the procession was drawn up: in front the bridegroom between his godfathers: next the bride between her godmothers, then the musicians with a harmonica, triangle and *estúrdia*, this last an instrument formed of a hollow reed with tin pans at either end; and at the rear the general company with the bearers of the wedding cakes. On the way they passed the tables, *ramos* and arches with offerings for the young couple.

It is the custom to place on the route which the young couple follow to the church, plates with food and bottles of wine on tables or even on the ground, those who make these offerings remaining hidden. The procession stops at each table, and it is the duty of the godfathers to take the presents, leaving in exchange wedding cakes, or when there are none left, presents of money.

After the religious ceremony, on reaching the church door, the wedded couple received the blessing of the godparents, after which all went to the bride's parents' house, following as far as possible a different route. The *alviçareiros*, those who first bring the news of the tying of the knot to the parents of the bridal pair, receive a wedding cake as a prize, wherefore there is no lack of small boys anxious to compete.

At the door of the bride's house the procession stopped, for the groom to entrust his wife to the care of his godmothers in the following terms:

"Lady godmothers: I entrust to you my wife, free and unhindered in order that you may return her to me when I come for her at an hour which I shall not yet say."

Not only the bride and her godmothers, but also the guests invited by her family, stayed in her house, the others proceeding to the house of the groom's father. It was time for dinner, and when this was over, all went on to the bride's house, into which they entered after the groom had declared that he had come to fetch his bride.

Then, after a round of dancing, the young couple went to their new home, before which they stood for a moment, while the husband, taking the keys from his purse, handed them to his wife, saying:

"My Lady wife, I give you the keys of our house that you may always open to me, at any hour, whether I come alone or with company. If corn-bread is ever lacking in this house we will eat maize-bread; and if there is only a sardine we will share it between us."

Taking the keys, the woman opened the door and entered the house with the right foot first. Then she turned to her man and said:

"Since my husband has given me the keys, O godfathers, godmothers and wedding guests, I pray you enter, for my husband has given his consent."[1]

Perhaps the most interesting moment in this narrative is when the bridegroom comes for the bride and is made to hunt for her himself. Clearly this is a relic of the ancient custom of marriage by capture, which was preserved in mime long after its practice had been abandoned and forgotten. Traces of it are found in other Portuguese customs, notably in a gypsy wedding

[1] Quoted in the *Guia de Portugal* (Biblioteca Nacional de Lisboa), II, pp. 511–12.

at Estremoz, described to me by Luis Chaves. Preceded by a band the cortège marched into the city. At the principal gate, they all stopped, and the bride suddenly made off across the market place, hotly pursued by the groom. The wedding guests shouted encouragement to the latter until he caught the girl, when they broke into applause. A dish was thrown up into the air and fell in fragments, and the couple were man and wife.

More faithful to the old tradition is the custom observed in the middle of last century at Jarmelo, where the bride's relatives used to try to prevent her from leaving the house. In the Serra de Monfurado, when the groom comes to fetch his betrothed, they try to foist every other girl in the house on him until at last they are compelled to give him the right one. Elsewhere the bride, on leaving her home, must weep and groan aloud to signify that she goes unwillingly.

In parts of Beira Alta a group of neighbours attempt to carry her off on the way to church, and it is only the stout resistance put up by two groups called the patrols (*patrulhas*) of the bride and bridegroom which serves to protect her. At Santa Vitória do Ameixial (Alentejo) the bride used, within living memory, to disguise herself after the religious ceremony, and hide from the groom, who had to find her before he could take her home.

In the mountainous country on the borders of Minho and Trás-os-Montes the ceremony of fetching the bride is more dignified though no less picturesque. In the Terras de Bouro and Barroso the groom and his relations go to the bride's house and a set dialogue ensues. The bride's eldest male relative asks:

"What seek you here?"

"A wife, honour and wealth," declares the suitor.

"Enter for you shall find all three," is the traditional reply. At Bouro it is more curious: "She has herded goats, she has leapt hedges, and if she has spiked herself on one of them, and you want her as she is, so do I give her to you."

To go to church the bridal pair are dressed in all their finery, the bride being hung with as much gold as she can muster. In some parts of the Minho they ride in an ox cart decorated with flowers and ribbons and hung with bells. The groom must go with his legs inside the cart and return with them dangling outside.

Another cart follows with the dowry and bridal chest. On reaching the church the oxen are unyoked, and the bridal pair themselves drag the heavy cart into the church to the spot in front of the altar where the ceremony is held.

On its way to, and especially from, church, the procession passes a number of "obstacles" such as the tables and arches mentioned in Senhor Eloi's narrative, all of which have some symbolic or even "sympathetic" significance. At Jarmelo the way is barred with ribbons or cords stretched across the road, which can only be passed on payment of a forfeit. Tables beside the road, at Trancoso for instance, are loaded with bread and wine or sweets, of which the happy pair partake. Sometimes they are set with a distaff, spindle, book and other similar objects, indicating the duties incumbent on married folk. A similar intention is to be discerned at Marco de Canavezes, where the bridal pair pass beneath three triumphal arches, on each of which different objects are hung. On the first are a distaff, paper and ink. The procession stops. For a moment or two the bride spins, while the groom writes a few words. On the second are a book and a pillow. The groom reads and the bride embroiders. On the last are hung a stocking and sword. The bride knits a few stitches and the groom unsheaths the sword.

At São Tiago da Cruz (Minho) a lemon and an apple are hung on an arch. The bride takes the first, the groom the second, and they then exchange them. The lemon is said to be given in order that their married life may retain its savour, and the apple that the bride may be exposed to no temptations. In parts of Estremadura a cake is hung on the arch. This cake, like those which figure in other parts of the ceremony, and like the sugared almonds, flour and confetti with which the departing couple are bombarded, stands for fruitfulness and plenty.[1] The

---

[1] At Parada de Infanções, in the district of Braganza, the bridal pair hold small cakes "phallic in form" which other guests try to snatch from them in the belief that if successful they will be married within the year.

ramos, mentioned above, are wooden staves, decorated with coloured paper and flowers, which are offered to the bridal pair, usually to the accompaniment of special songs, e.g.:

Aqui tem este raminho	Take then this little *ramo*
Da minha mão delicada.	From my delicate hand.
Se algum dia foi solteira	If once you were a maid
Agora está casada.	You are now a wedded wife.

In general the proceedings are enlivened by plenty of noise, and by much firing of muskets and setting off of rockets, intended in their origin to scare away the demons at this critical moment in the fortunes of the bridal pair. The latter derive additional powers of resistance from the copious wedding feast. At a wedding at Penedones (Barroso) in 1874, four calves and three pipes of wine were consumed and no less than four hundred kilogrammes of gunpowder were expended. If the bride is not spotless there are no *ramos*, no muskets and no rockets. On the way home, or during the banquet, special songs are sung, congratulating the happy couple, and giving them good advice for the future:

Fostes hoje á igreja,	You went to-day to church,
Minha salvinha de prata;	My little silver sage;
Fostes dar um nó tão cego	You went to tie a knot so tight
Que só a morte o desata.	That only death can loosen it.
Ó que lindo Sacramento	How lovely a sacrament
Fizeram estes senhores;	Have this fine pair taken;
Deus no céu lhes bote as ben-	May God in Heaven shower
cões	blessings on them
E nós cá na terra as flores.	As we on earth shower flowers.
Essa rosa, Senhor noivo,	Yonder rose, my Lord Bride-
	groom,
Inda ontem era botão,	Was only yesterday a bud,
Trate dela como sua,	Treat her as you would your own
Meta-a no seu coração.	And place her in your heart.

The shirt or shift worn at a wedding is often treasured up till death, for it is thought good to wear it on that last journey from which there is no return. This is not, indeed, the only example in Portugal of the close parallelism which exists between the rites of marriage and death. According to a superstition, common in varying forms through the country, it is in a bridal pair the one before whom the lights burn weakest during the ceremony, or who is the first to enter the bridal bed or to put out the light on the wedding night, who will be the first to die. Yet another link, reported to me from the Algarve, is the tradition which leads aspirants to the hand of a widower to keep the vigil over the body of his dead wife.

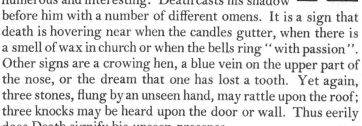

Portuguese superstitions regarding death are numerous and interesting. Death casts his shadow before him with a number of different omens. It is a sign that death is hovering near when the candles gutter, when there is a smell of wax in church or when the bells ring "with passion". Other signs are a crowing hen, a blue vein on the upper part of the nose, or the dream that one has lost a tooth. Yet again, three stones, flung by an unseen hand, may rattle upon the roof; three knocks may be heard upon the door or wall. Thus eerily does Death signify his unseen presence.

When the death agony begins, the relatives of the dying man whisper messages in his ear for those who have preceded him into the other world. Such messages may also be given immediately after death, or a letter to the beyond may be slipped into his pocket. According to an old custom, not yet extinct in the North, a lighted candle is often placed in the trembling hand of a dying man. King João II is recorded to have expired *candeia na mão*. The last moments of the sufferer may also be rendered painful by the ministrations of *despenadeiras*, women who, pressing upon his breast, seek by expelling the last breath from the lungs, to facilitate the departure of the soul from the body.

The soul flies away in the form of a moth "white as a snowflake" which hovers for a moment upon the lips and is gone. Thereupon the bells of the church are tolled, three peals for a man and only two for a woman. Omens are taken from the expression of the features. Should they still bear the mien of the departed, the soul has reached a safe haven. Into the bodies of small children or of those who have died in the odour of sanctity, pins are stuck in the belief that by a current of sympathy those who implanted them will, when their turn comes, be taken into Paradise.

The corpse is washed, shaved and clothed in its best for burial. Should the stiffened limbs hinder these operations, the deceased is called upon by name to bend an arm or leg. It is said that he complies. At Vinhais a custom now obsolete ordained that when the body was removed a member of the family should take its place in the bed.

Death rites reflect, in Portugal as elsewhere, three ancient and primitive conceptions: that a corpse is ceremonially unclean and this uncleanness extends not only to the dead man's belongings but also to members of his family; that the dead envy and persecute the living; and that the presence of death endangers the lives of those around.[1]

The first of these notions, which is probably the oldest and most general, is reflected in many Portuguese usages. A dead man's mattress is burned, as among the Basques, and it is opined that if the smoke mounts straight upwards the soul has gone to Heaven; if it curls to right or left, the soul has gone to Purgatory or Hell. In the *terras de Barroso* the women of the bereaved family cut their hair and the men their beards. Ornaments are removed or covered, and the bells are taken from the cattle. A widow or widower retains a dirty shirt or goes unshaven for a month and at the end of this period attends a purificatory Mass. The lamp which burned in the death chamber must on no account be used again. The very house is regarded as unclean. All the water is thrown out, for it is thought that the soul of the deceased bathes in it. Food is sent in by the neighbours, sometimes for as long as a year. During this time no cooking must be done within the house, for fear of "frying the soul of the deceased".

A belief connected with the straw mattress links the idea of uncleanness with that of fear of the dead. The mattress is burned, it is said, in order that the spirit of the dead may not

---

[1] In a paper entitled "Whence Comes the Dread of Ghosts and Evil Spirits?" read before the Folk Lore Society and printed in *Folk Lore* for June 1933, Miss M. E. Durham attempts to explain the fear of ghosts with its attendant customs and taboos as reflecting a dim perception of the nature of infectious disease.

Plate VII

AN OLD PEASANT

[*facing p. 96*

Plate VIII

ENTRADA DE TOIROS AT VILA FRANCA DA XIRA

CARNIVAL COSTUMES IN LISBON

return to earth. Attempts to scare away the ghosts of the dead, common in primitive communities, are rarely found in Portugal, where, indeed, the *almas do purgatório* are generally regarded as harmless. But steps are taken to prevent their return by facilitating their departure, helping them on their journey, and generally propitiating them with offerings of food or money. To help the soul to escape, besides the attentions of the *despenadeiras*, it is usual to open the window of the dead man's house, or, strangely enough, his neighbour's. In one part of the Minho, when a funeral procession reaches a parish boundary marked by a bridge, the male mourners throw a handful of fine sand into the stream, stopping their ears in order not to hear the splash and saying: "So-and-so, may as many angels guide thee to Paradise as grains of sand have struck the water." This clearly aims at preventing the soul of the deceased from ever re-entering the bounds of the parish.

At Santa Isabel do Monte (Minho) it was till recently the custom to place within the coffin a small jar of water, a crust of bread dipped in wine and a small coin. At Sobrado a similar offering is made, consisting of a threaded needle, a reel of thread, a thimble and a pair of scissors in order that "the dead · man may mend his clothing in the other world". The explanation given for such offerings varies from place to place and affords an interesting insight into folk-belief. A garland of roses will, it is thought, commend the soul to God. A crust of bread will protect it from evil vapours (*cortar os maus ares*).

In many districts the offering of bread is connected with the belief that the soul must either pass through a little hole at Santiago de Compostela,[1] through which every mortal must crawl either in life or in death, or cross the "Bridge of Santiago" which is pictured either as a real bridge at the great Galician place of pilgrimage, or as that vaster and more widely known "Bridge of St James", the Milky Way. Should the soul not complete this task between the death and burial of the body, should the body, in fact, be buried in the absence of the soul, the latter will be lost.

[1] There is little doubt that the great mediaeval cult of St James of Compostela was founded on a pre-existent pagan cult.

The crust of bread is often regarded as the offering required to placate a lion which guards the bridge and will not let the soul pass until it has swallowed the bread.

The bridge, moreover, is open, and the Devil lies in wait below for the souls who fall through its interstices. The coin is usually regarded as a tribute to the Devil or to São Hilario who by some queer misconception is often confused with him. For this reason, the coin chosen must not be one graven, like so many Portuguese coins, with the Cross of Christ. According to a belief not easy to explain no woman can attain Paradise a virgin. If she departs from earth in this state, she is deflowered by St Hilary. Indeed, peasants have been heard to exclaim, on the death of a young girl, "St Hilary is in luck again". In certain villages the coin in the coffin is regarded as the price for which this highly peculiar saint will relinquish his *droit du seigneur*.

This guardian of the other world, lion, Devil or Saint, who lets none pass save on payment of a tribute, has a distinctly classical flavour. Indeed the grim boatman of the Styx looms large behind the belief recorded by Leite de Vasconcelos, that the coin is for the purpose of "crossing in the boat of St James"

Offerings on behalf of the dead are not always enclosed in the coffin. Sometimes they are consumed by the relations, distributed to the poor or handed over to the Church. For centuries the latter was unsuccessful in stamping out feasting on graves (e.g. Oporto *Constituições* of 1687) but, though still remembered at Carrazedo de Bouro, this custom has now been generally replaced by a funeral banquet the original purpose of which is clearly discerned in its name of *bodo* (Lat. *votum*).

On the first Sunday after a death the bereaved family send to church an offering called an *obrada*, of corn, codfish, and wine, which are placed on the altar. After Mass, an *obradoiro* is held, consisting of prayers and responses for the soul of the deceased, after which the offering is consumed by all present or given to the poor. In Beira Alta bread is distributed to the poor at the church door on the day of the funeral. In some places, in spite of ecclesiastical prohibition, an *agasalho* of bread and wine is distributed to members of the religious fraternities, special provision for which is found in some eighteenth-century wills. A similar offering of bread, wine and fish used to be the traditional reward of the *carpideiras* (professional mourning women) who, in spite of ecclesiastical censure, survived fifty years ago in the wild Soajo country, and may still be heard to-day round Braganza.

Just as the ritual connected with the mattress linked the idea of purification with that of preventing the return of the departed soul, it is also connected in at least one locality with the fear that Death, once he has made his appearance, will carry off other victims. In the Douro it is believed that if a dead man's mattress is not turned and set with the head where the feet once were, another member of the family will die. A similar fear is reflected in a number of taboos observed all over the country, the most usual of which dictates that no living being shall remain asleep or even lying down when a funeral leaves the house or passes the door. Children, in particular, are roused when a funeral passes. The very animals are made to stand up and eat when it is about to take its departure. Those who neglect to observe these precautions will surely be taken soon afterwards, and so will anyone who watches a funeral out of sight, at whose door it stops, who stands on the shady side of the road during its passage or who accidentally extinguishes the light that has burned in the death chamber.

This chapter may fittingly conclude with a very curious custom from the Azores. Anyone who has suffered injury from a man during his lifetime will follow the funeral on his enemy's death, strewing *tremoço*, barley and salt, and saying:

Quando este tremoço nascer	When this lupin takes root
Esta cevada enrelvar	And this barley shows ear
Este sal temperar,	And this salt seasons,
Seja quando me volte a inquietar.	Not till then may you plague me again.

CHAPTER V

SPRING MAGIC

In the early days of Christianity, the Church strengthened its hold over the newly converted by admitting into its ritual many ceremonies of older faiths. By consecrating these, as it were, and endowing them with new significance, it sought to harness to the cause of Christianity the forces of habit and tradition.

So tenacious was the hold upon Europe of the ancient modes of thought and ritual that the Church had to wage a long and bitter struggle to suppress the very essence of paganism. One has only to think of the countless anathemata pronounced by ecclesiastical councils during the first millennium after Christ against openly pagan practices, and of the abuses which, in the Middle Ages, were still to be found within the very orbit of the Church. What wonder then that the latter was willing to transact with its adversaries over non-essentials?

Of all pagan observances, it was the festivals of the Calendar, strengthened in tradition by the regularity of their recurrence, which proved the hardest to eradicate. To provide them with a new explanation and justification, the Church chose to celebrate the birth of Jesus Christ at the winter solstice, thus transferring the devotion of the heathen from the sun to Him who was called the Sun of Righteousness, and to assimilate the Easter festival of death and resurrection to a pre-existent festival of the same character falling at the same season. Similarly, by superimposing the feast of St John the Baptist on a Midsummer festival of water, and that of All Souls on an autumnal festival of the dead, they were able to graft the new faith on the old tree of paganism.

Of these ceremonies Sir James Frazer writes in one of the most sonorous passages of the *Golden Bough* that they

are, or were in their origin, magical rites intended to ensure the revival of nature in Spring. The means by which they were supposed to effect this end were imitation and sympathy. Led astray by his ignorance of the true causes of things, primitive man believed that in order to produce the great phenomena of nature on which his life depended he had only to imitate them, and that immediately by a secret sympathy or mystic influence the little drama which he acted in forest glade or mountain-dell, on desert plain or wind-swept shore, would be taken up and repeated by mightier actors on a vaster stage. He fancied that by masquerading in leaves and flowers he helped the bare earth to clothe herself with verdure, and that by playing the death and burial of winter he drove that gloomy season away, and made smooth the path for the footsteps of returning Spring.... After all, magical ceremonies are nothing but experiments which have failed and which continue to be repeated merely because... the operator is unaware of their failure. With the advance of knowledge these ceremonies either cease to be performed altogether, or are kept up from force of habit long after the intention with which they were instituted had been forgotten. Thus fallen from their high estate, no longer regarded as solemn rites on the punctual performance of which the welfare and even the life of the community depend, they sink gradually to the level of simple pageants, mummeries and pastimes, till in the final stage of degeneration they are wholly abandoned by older people, and from having once been the most serious occupation of the sage, become at last the idle sport of children.[1]

In Portugal, as in the rest of Europe, the original purpose of such ceremonies is seldom remembered, and those who enact them are generally ignorant of the real meaning of their actions. To them these are either a form of recreation and entertainment, marking the passage of time and relieving its monotony; or a tradition carried on by the sheer momentum of ingrained habit; or even, in some cases, just an excuse for children to beg presents of food and money.

Such are the wassailing songs, known as *janeiras* or *reis*, with which the New Year is ushered in by untutored childish voices.

The custom of singing carols on the eve of important festivals was practised in Ancient Greece, as it still is throughout modern Europe. In the British Isles we have our wassailing and gooding songs, our guisers, pace-eggers and wren boys.

[1] *The Golden Bough*, Abridged ed., pp. 320, 322.

The substance of the songs is always much the same. The children go round all the houses wishing their inmates good fortune and receiving in return presents of food and money. If they are made welcome they pay compliments to the master and mistress of the house, but they assail with insults any who turn a deaf ear. It becomes clear that only those who reward the singers can participate in the good luck which, implicitly and often explicitly, they claim to bring into the houses. In many parts of Europe they carry the "luck" with them in concrete form, in the shape of greenery, a doll, or, as in Navarre on New Year's Eve, of fresh rain or well water. Of this there seems to be no trace in Portugal except at Foz do Dão, where the leader of the waits carries a light which must not be allowed to go out.

It is easy to understand why this ceremonial peddling of the "luck" should take place on New Year's Eve, for the first of January is regarded as being of special omen for the ensuing year. The peasants of the Minho believe that as they act on this day so they will continue to act throughout the year. Thus, they are disinclined to pay for anything on New Year's Day, in the belief that if they do so they will continue to part with money till the year's end. As in the British Isles, moreover, the belief is current that the first twelve days of January indicate the weather of the corresponding twelve months of the year.

Documentary evidence of the *janeiras* is found as far back as 1385. Unfortunately we do not possess the verses of the folk of Évora who, on New Year's Eve, 1638, congregated before the house in which the Conde de Linhares was lodged "in order to sing to him certain benedictions and rogations, a custom of our ancestors, which, under the name of *janeiras*, they intoned at the doors of their dearest friends". But these cannot have been very different from those which the masked *janeireiros* still sing at Loulé when they receive their traditional gifts of sausage, onions, chestnuts, apples and wine:

Long live the Lady of this house, who is a branch of silver willow,
Her body is as snow, and her soul the soul of a Saint.

or, in the Serra de Monchique:

Lady Mistress of this house, little branch of green willow,
There, at the foot of your bed, rises the sun and sets the moon.

The next day, children knock at the doors with little baskets, shouting: "*Janeiras, janeiras!* Our Lord will give good crops."

At Ponte de Lima, these two verses were sung to my friends the Count and Countess Aurora:

Viva lá o Senhor Conde	Long live my Lord Count
Colarinho engomado,	In his starched collar,
No meio da sua sala	Standing in his hall
Parece um doutor formado.	He looks like a Doctor of Law.
Viva lá a Senhora Condessa	Long live my Lady Countess
Vestidinha de cambraia,	Dressed all in lawn,
Quando se poe á janela	When she appears at the window
Alumia toda a praia.	She brightens all the shore.

A less agreeable parting-shot, reserved for those who have given nothing, is:

> This house stinks of balm; there dwells here some corpse;
> This house stinks of pitch; there dwells here some Jew.

The *reis* are nothing but the "pagan *janeiras*" sanctified by transference to Twelfth Night. The change is noticeable principally in the words, and very often the Three Kings of Orient themselves accompany the carollers. The strictly religious part of the song is seldom of popular inspiration and betrays erudite clerical influences. But the compliments and begging verses are never long in following:

> Long live Senhor X., a gold pin at his breast,
> When he passes the girls he winks his right eye at them.

> Long live Senhora X., the flower of the peony,
> She is as fair as the sun and as clear as the day.

> Let them give us the *reis*, if they are going to give them,
> We have come from far away, near the coast of the sea.

> These bran-beards have nothing to give us,
> All they have is a larder that the rats befoul.

This last verse, among others, closely resembles those sung in the Basque provinces and elsewhere.

Twelfth Night, or more rarely the "feast of Stephen", is the occasion on which, in the Braganza district, a curious festival called the *festa dos rapazes* is held by the young unmarried men over sixteen. They hire a piper and sup off a calf which they have bought between them. With masked faces and raucous cries they then go the round of the village. Some of the masks represent the heads of animals such as bulls and goats. Their leader, known as the *Juiz da Festa*, is unmasked but wears a tall crown, recalling the old custom, recorded in documents from this same district, of electing an "emperor" on St Stephen's Day. The young men conclude with an improvised play or satirical verses sung from a platform, holding up to ridicule such of their fellow villagers as have suffered some slight misfortune during the past year, such as a toss from a bull or the loss of an escaped pig. The performance, which is not unlike our own Kirtlington Lamb Ale, is thoroughly pagan and mediaeval in character, and recalls the sixth-century edict excommunicating "any who on the kalends of January clothe themselves with the skins of cattle or carry about the heads of animals".

The election of an "emperor", dating from the time when the functions of king and priest were united in one person, and connected in all probability with the ancient tradition of the "King for a Day", has been transferred in other parts of Portugal to Whit Sunday. At the *festa do Imperio*, founded at Alenquer by the Holy Queen Isabel, not only an emperor but two kings were crowned in church with great pomp and ceremony, and the former had the right to free a condemned prisoner.

The *Festa do Imperador* at Colares has long since fallen into disuse, but that of Alcabideche just behind Cascais and almost within sight of Lisbon survived till the Republic and is still remembered in the neighbourhood. It was held not annually but at fairly frequent intervals in fulfilment of a vow by some well-to-do farmer, who paid all the expenses of the festival and whose son, who had to be about sixteen years old and un-married, played the part of the "emperor". The young man was arrayed in church vestments and escorted by four attendants into the church, where, with the greatest solemnity, he was crowned with a sort of mitre ("Oh the hat! So high and like silver!"). He then sat on a throne by the altar with a youth on either side "to keep the flies off him", and finally, while all knelt, the parish priest did him obeisance. After the ceremony, there was a *bodo* (banquet) at the end of which, his brief reign ended, the "emperor" doffed his robes and, like the boy bishop of Berden (Essex), became a rustic once more.

In February or early March there falls a festival which, in all parts of Southern Europe, is associated with interesting survivals. This is the Carnival which begins some weeks before Lent and reaches its apogee in the three days preceding Ash Wednesday. Although we may reject the theory that the festival was originally connected with the opening of navigation each spring, and the name  derived from *carrus navalis* (the ship of the goddess of navigation, Isis in the Mediterranean, Nehalennia in Northern Europe), there is no doubt that the Carnival has inherited many features from earlier festivals such as the Greek Kronia and the Roman Saturnalia. It may be objected that the latter festival was held in December, and did not therefore coincide in season with the mediaeval and modern Carnival. Originally, however, the latter lasted from Christmas till the beginning of Lent, and there is little doubt that it was to some extent grafted on to the earlier festivals, although it seems to have assimilated a wide variety of those ritual elements which were scarcely respectable enough to earn inclusion in the church calendar.

The Saturnalia was the most famous of a number of periods of licence, observed throughout the ancient world, "when the customary restraints of law and morality are thrown aside, when the whole population give themselves up to extravagant mirth, and jollity, and when the darker passions find a vent which would never be allowed them in the more staid and sober course of ordinary life".[1] If this description is a little too highly coloured to be applied to the Carnival of to-day, it would have been by no means out of place in mediaeval, and indeed in relatively modern, times.

[1] *The Golden Bough*, Abridged ed., p. 583.

A hundred years ago the Lisbon Carnival was an orgy of coarse jokes and rough horseplay, leading a foreign observer to exclaim that "the Portuguese lost their wits on Carnival Sunday, and only recovered them on Ash Wednesday". "In the middling classes", wrote A. P. D. G. in 1826, "the frolics of the Carnival consist in throwing hair powder and water in each others' faces and over their clothes; and pelting the passengers in the streets with oranges, eggs and many other missiles besides throwing buckets of water on them."[1] In a later passage the same writer mentions small india-rubber squirts, stuffed gloves smeared with grease and chimney-black, dust, and wax water-bombs, more faithful successors to the flour and filth of Ancient Rome than are the petals and confetti of a Riviera battle of flowers. People organised "assaults" on the houses of their friends, and it is on record that in one such *assalto* on a popular actress living in the Rocio Square no fewer than six hundred eggs were used as missiles.

If to-day Carnival is almost restricted in the capital to the wearing of fancy dress and to processions of decorated cars and carriages (features of the Saturnalia no less than those already mentioned), it must not be thought that this change has been brought about in a moment, or that attempts to mitigate the abuses of Carnival are a new thing. In 1604 the use of squirts and the pleasant habit of hitting people with rotten oranges was prohibited by order. In 1689 another order formally forbade the wearing of masks, under cover of which countless crimes had been committed and private revenges satisfied. In the reign of Pedro II a breach of this order was punishable by four years' exile in Angola. But laws are made to be broken, and it is only within living memory that Carnival has been brought within bounds, while the odious habit of attacking ladies with scent or ether sprays was only suppressed a few years ago.

Even to-day private houses are sometimes the scene of parties at which it would be more convenient, if less correct, to wear football clothes, flour and eggs combining to form

[1] A. P. D. G., *Sketches of Portuguese Life, Manners, etc.*, 1826.

streaming omelettes on the persons of the guests. The telephone, moreover, has given a new lease of life to another deep-rooted tradition which was fast dying out in the city, to wit the *pulha*. This name is applied to an obscene jest or insult which in less fastidious days was addressed openly by one sex to the other, but which to-day will only greet the guileless who unsuspectingly place their ear to the receiver. It may be asked whether it is not as true to-day as, if one is to believe the *Diario do Governo* of February 13th, 1836, it was a century ago, that "well-educated families and some highly placed ones, are those who most distinguish themselves in these excesses".

Although it is on the wane, Carnival survives in the provinces in a more robust and less sophisticated form, with battles of flour, rotten eggs, oranges and wax water-bombs, with infernal serenades and cacophonic "rough music", with soot-blackened faces, with tilting on horseback at bags of filth and with *pulhas* which disdain the secrecy of the telephone.

At Espariz (near Coimbra), for instance, voices are heard among the hills shouting through improvised megaphones the most outrageous gibes at any husbands who are thought to have sprouted horns during the year. This custom is called the *corredela do Entrudo*, and is typical of the North where, as Lent draws near, the countryside resounds with insults and obscenities.

The reign of licence used to reach its climax in the *procissão dos fogareus* at Braga, which by some strange aberration was suffered to intrude its ribaldry into the solemn devotions of Holy Week. The religious procession held on Maundy Thursday was preceded round the town by a cortège of torch-bearers masked as penitents or Roman soldiers who exercised the traditional privilege of shouting aloud the secret sins of the inhabitants. No calumny was too vile or scurrilous, no charge too false or too true, to be proclaimed in this manner to all and sundry.

The exchange of insults and obscenities can doubtless be attributed partially to the licence prevailing at this period. But the obscene element merits a little closer attention. Not only may it prove to be less puerile than at first appears, but it may provide the key to one of the most important aspects of Carnival.

Frazer shows that "annual expulsions of demons, witches or evil influences appear to have been common among the heathen of Europe...and this public and periodic expulsion of devils is commonly preceded or followed by a period of general licence, during which the ordinary restraints of society are thrown aside, and all offences, short of the gravest, are allowed to pass unpunished". It is easy to understand that the ordinary rules of conduct should have been relaxed immediately before or after a general cleansing of evils, as exemplified in the licence prevailing at Carnival time.

The exchange of obscenities, however, also recalls the Eleusinian mysteries during which the use of scurrilous language and the breaking of ribald jests was not merely tolerated but positively enjoined on the participants. There is every reason to believe that the Eleusinian mysteries were a highly developed agricultural fertility cult; and, in accordance with the principles of sympathetic magic, the participants in such a cult may well have thought that they were promoting fertility by the open mention of that which causes it.

The frequent recurrence of the obscene element in Carnival festivities would seem therefore to confirm that at least one element in the latter is drawn from an ancient fertility cult, such as was the Saturnalia. There is nothing inherently improbable in this suggestion, for Carnival occurs in early spring, the most critical time for both animal and vegetable fertility.

This is strikingly borne out by a chance discovery which I made one day while playing golf at Estoril, normally an unpromising field for folkloric investigation. Among the pine woods behind the course I heard the sound of someone playing monotonously, yet with curious insistence, on a primitive reed pipe with only two or three notes. At first I paid no attention, thinking that it was some child with a toy whistle. When this succession of little trills and flourishes had continued without interruption for more than an hour, I asked my caddie, a peasant boy, what this music might be. His reply was unexpectedly illuminating: *São os rapazes que chamam*

Entrudo ("It is the boys calling *Entrudo*"—a name given indiscriminately to Carnival or its personification). The connection, clear enough in itself, between this summoning of *Entrudo*, and a spring magic intended to hasten the advent of spring, and to call up the tender shoots from the buried seed, was confirmed by a further explanation that "they do it because they want winter to come to an end".

It has been shown to be one of the principal tenets of sympathetic magic that the most reliable way of producing an effect is to imitate it. It is in this light that a mumming-play should be viewed, upon which it was our good fortune to chance in the village of Cadriceira near Torres Vedras on Carnival Sunday in 1932.

We first saw the mummers marching along the main road, preceded by their musicians and followed by a little knot of excited children. The two musicians wore knee breeches, white stockings and green and red stocking caps. One of them played a flute, and the other the oddest percussion instrument I have ever seen, an empty earthenware jar, which he struck with a small wicker fan of the type used to fan the embers of a charcoal stove. They were accompanied by an incongruous figure in a straw hat, carrying a stick with a round top, broad and flat, on which miniature platform a couple of dolls, dressed as peasant man and woman, jigged, by some ingenious mechanism, in lifelike imitation of the *bailarico*.

A group of costumed young men followed. At first sight there seemed to be nothing to distinguish them from a *cégada*, one of the bands of masqueraders who at Carnival time dance and play their way round the city and countryside. One of their number, however, carried an object which caught our attention and seemed to hint at the possibility of something more elaborate. This was a pair of thin posts, some six feet high, to which was sewn a piece of printed calico. With some curiosity we followed the little band off the road into the village.

On reaching an open space near the church, they stopped, and the purpose of the posts became apparent. These were planted in the ground a yard or so apart, the strip of calico forming the backcloth of an imaginary stage, before which the villagers grouped themselves. The mummers retired behind it, and then, each in his turn, made their entry through the "curtain". The first to appear wore a battered top hat and frayed frock coat and carried a black bag in his hand. Singing his lines unaccompanied he announced that he was a doctor, famous throughout Europe and able to cure any disease. Two more characters now entered, a peasant and his wife. The first wore an old-fashioned broad felt hat and side-whiskers, and carried a striped saddle-cloth over his shoulder. His lady, in check dress and cunningly devised 'kerchief and veil, acted her part so well that our suspicions of her real sex were not confirmed until a pleasant tenor voice betrayed her.

The dialogue which followed, and indeed the whole play, was sung in quatrains to the same tune. The wife described her symptoms at great length, and the husband eloquently implored the doctor's aid. Presently there entered a new character wearing a white overall and bringing a chair on which the lady sat down with heart-rending groans to be examined by the doctor. A sudden surmise leapt into my mind Would the subject of this Portuguese mumming-play be the traditional theme of such performances from Roumania to the Yorkshire dales? It was. A moment later, with a deft gesture the doctor slipped his hand under the chair and whipped out— a celluloid doll. The audience guffawed, shrieked or blushed scarlet at what, no doubt, they had witnessed at every Carnival of their lives.

The play continued, verse succeeding verse, but for us the climax had already been reached, and we took little note o the rest until the end, when all the characters, having declared themselves to be young men from Vila Franca do Rosario, wished their audience good luck, concluding:

Meus Senhores pedimos disculpa
São versos do Carnevale.

It was impossible to grudge them the pardon for which they thus asked. Their play had been no senseless piece of bawdiness. Its rabelaisian quality was, as they suggested, excused by the Carnival. The magic birth of ancient ritual had been accomplished in mime, and by this sympathetic magic the fertility of the village and all that went to its sustenance had been assured for the coming year.

The actors, of course, were not consciously aware of this. They had in fact written the lines of their play themselves that very year. But their theme had been determined by an ancient tradition, its roots lost in the mists of the past. And it is just such a performance as theirs, the mock birth of a child on a harvest field in West Prussia, which, in Frazer's opinion, 'points to an older practice of performing among the sprouting crops in spring or the stubble in autumn, one of those real or mimic acts of procreation by which...primitive man often seeks to infuse his own vigorous life into the languid or decaying energies of nature".[1]

It is difficult not to see a mimic act of procreation in a custom till practised at Cercal near Valença do Minho under the name of *rebolada*. In May or June, young men and girls go into the flax fields just before the flax is pulled, and roll together on the ground in couples. At Santo Tirso the same rite is performed by one couple chosen by lot and is known as *talhar a amisa* (cutting the shirt).[2]

On our way home that Carnival Sunday we came upon a play of a very different sort in a village much nearer Lisbon. A young boy and a white-bearded old man were clinging pathetically to each other, while another actor, taking the part of a hard-hearted landlord, was turning them out into the street.

[1] *The Golden Bough*, Abridged ed., p. 421.
[2] There are indications that in the Elvas district a similar act was performed the end of the olive gathering as late as 1863—and not in mime.

The dialogue of this moving melodrama was sung throughout to a particularly lugubrious *fado* tune, and was interspersed with phrases of an exquisite bathos, such as: *Ai que triste situação!* ("Oh, what a sad situation") and *Isto é uma desgraça!* ("This is indeed a misfortune").

The issue was appropriately dramatic. A sudden thought strikes the villain. "Were you ever married?", he asks the father of the fifteen-year-old boy. "Alas, I cannot remember", replies the old man. The villain then discovers himself to be a long-lost child, and the two brothers, thus reunited, lead off their *pai tuberculoso* to a new life of happiness together. The audience were much moved. And, degenerate as it was, the performance showed that even so near Lisbon the feeling lingers on that Carnival is a time to go a-mumming.

This *fado* scene is many degrees removed from spring magic. But there still remains to be considered a custom, associated with the Carnival throughout Southern Europe, which links it with the Saturnalia and establishes firmly its connection with fertility rites, however remote it may at first sight appear to be from the latter.

In Rome a mock king of the Saturnalia was elected each year, who, in earlier days and remoter parts of the empire, was actually put to death at the end of his brief reign. The counterpart of this human king may be recognised in the puppet or figurant who so often personifies the Carnival (or the licence or gluttony associated with it), and who on Shrove Tuesday or Ash Wednesday is either executed in mime or publicly burned to the feigned grief and genuine delight of the populace.

On the outskirts of Lisbon and in the very city itself, boys or young men go round in the afternoon or evening of Shrove Tuesday, carrying on a stretcher one of their number with his face blackened, and begging contributions for the *enterro do Entrudo* (burial of Carnival). A common custom throughout the provinces on Ash Wednesday is the "funeral" with mock lamentations, of a figure named *Bacalhau* (after the dried cod which is the usual meagre fare of the Portuguese during Lent) who usually meets his end on a bonfire. Among the dense corkwoods of the Serra de Monfurado (between Évora and

Montemór o Novo) a more elaborate ritual takes place. The last four Sundays before Lent are dedicated to groups of boys and girls who call themselves *amigos, amigas, compadres* and *comadres*. Each group, when its Sunday approaches, constructs with great secrecy a straw effigy of the opposite sex, and nominates two of the group to be its parents. On the appropriate Sunday these set up a great wailing and lamentation, proclaiming that their child is dead. The last puppet, the male one of the *comadres*, is burned with great ceremony at the end of Carnival.

In Beira Alta, on the two Thursdays before Lent, a similar custom is observed. The women make a puppet called a *compadre*, which, since they are not allowed in the streets, they hang from a window. The men make a *comadre* and, with their faces masked, carry her about the streets impaled on a long pole.

At Estremoz, on the *quinta feira de compadres*, the boys carry silken banners and deck themselves out with flowers, ribbons and handkerchiefs. The girls parody them with banners of rags and parade about with old bones, kettles, brushes and mats. On the following Thursday the rôles are reversed, and on both occasions the two groups come into conflict and try to carry off each others' trophies.

The masked or blackened faces and the general air of secrecy are familiar features of ritual survivals, and those who are familiar with the studies of Mannhardt and Frazer will have no difficulty in recognising, in these puppets or figurants, representatives of the spirit of vegetation who is annually put to death in order that the crops may prosper. For the benefit of those who have not read the *Golden Bough* Frazer's theories on this subject may be briefly summarised.

It has already been explained how in the gradually awakening intellect of primitive man animism led first to polytheism and eventually to anthropomorphism: or, to take a particular instance, how the idea that every tree or cornfield is the abode of an individual spirit gave way to the conception of a general tree or corn spirit, which then began "to change his shape and assume the body of a man in virtue of a general tendency of early thought to clothe all abstract spiritual beings in concrete human form".

It is a far cry from present modes of religious thought to the custom of killing a beneficent spirit or god, even vicariously. Nevertheless, there is little or no doubt that such a custom prevailed widely in a bygone age, and, once the principles of sympathetic magic are granted, it admits of a logical interpretation. Primitive man believed that if his gods, or their  human representatives, were allowed to grow old and to fail in strength, this circumstance would be fraught with disaster for the community. The custom accordingly grew up of killing the man-god when his strength began to fail or at a fixed interval which in some cases was annual. This sacrifice was in no way intended as an act of propitiation, but was a practical measure destined to assure that his divinity and strength were transferred to a vigorous successor before it had been seriously impaired by the threatened decay. In other words, the killing of the god in the person of his human incarnation was merely a necessary step to his revival or resurrection in a more youthful and vigorous form.

Now, as Frazer points out, "of the annual phenomena of nature there is none which suggests so obviously the idea of death and resurrection as the disappearance and reappearance of vegetation in autumn and spring". It is therefore hardly surprising to find evidences of the custom, mimic in the present day but gruesomely real in the age when it expressed a living belief, of annually killing a human representative of the spirit of vegetation associated with the tree or the corn.

The analogy between ancient practice and modern pantomime is often striking in the extreme. From the myth of Lityerses as sung by the Phrygian harvesters, it can be deduced that it was once the custom to seize a stranger who happened to pass a cornfield and to put him to death as a human embodiment of the corn spirit. The mythical Lityerses used to regale the passing stranger with food and drink, then take him to the cornfields on the banks of the Meander and compel him to reap with him. Lastly he would wrap the stranger in a sheaf, cut off his head with a sickle and carry away his body swathed in corn stalks. At Carnival time in the Vinhais district of Tras-os-Montes, young men disguised as Death or as the Devil seize any passing stranger and make him kneel on the bare earth while the assembled multitude cry "Death!"

In Central Europe there is a widespread custom of carrying out from the village a puppet called Death, which is generally followed by a ceremony (or at least a profession) of bringing back Summer, Spring or Life, often in the form of leafy branches. The second half of this ceremony, so far as the Portuguese Carnival is concerned, is preserved by bands of young men or girls in various parts of the country who dance with garlanded wooden hoops, like the *arcos grandes* of the Spanish Basque sword dancers. It is significant that little earthenware statuettes of these dancers made at Estremoz used to be known by the name of *Primaveras*.

Folk-lore, though a science, cannot claim to be an exact one, and the connection between ancient practice and modern survival can never be irrefutably proved. But in almost every European country a vast mass of evidence lends probability to the conjecture that the names, Carnival, Death, etc., are comparatively late and inadequate expressions for the beings personified or embodied in these customs. As human thought moved away from the conception of killing the god, even with the unexceptionable motive of perpetuating the divine energies in the fullness of youthful vigour, the effigy thus slain came to be identified with persons or things regarded with general aversion, such as Bacalhau, Judas Iscariot or Guy Fawkes. The custom of burning a straw effigy of Judas Iscariot on Easter Eve, which used to be observed in parts of Germany, is common throughout Portugal, even though the traitor is often clad in the blue overalls of a mechanic and hung from telegraph wires. At Monchique this ceremony is performed with great gravity in the presence of the parish priest. The effigy is stuffed with rockets and fireworks which, on exploding, release a black cat previously sewn up inside. The ashes of the German Judas were afterwards kept and planted in the fields on May 1st to preserve the wheat from blight and mildew; and the ashes of other effigies, like the flesh and ashes of real sacrificial victims among primitive tribes, are believed to be endowed with a magical or physical power of fertilising the land. Although I have not been able to trace any such belief in Portugal, the weight of evidence goes to show that all these customs were originally intended for the promotion of fertility. Even to-day the Carnival is openly as well as implicitly associated with this idea. In the Minho it is the custom for a farmer to go out into his fields on Shrove Tuesday and shout to his neighbour: "Millet for us and maize for you"; and then to blow as loud as he can on a horn. The neighbour replies in the same words, adding: "May you be punished." Both cry and curse are then passed on to other farms.

The black cat of Monchique recalls another similar cruel and barbaric feature included in the Carnival festivities at Arganil, and I believe at other places also. A cat was placed in an earthenware jar, which was fastened at the top of a tall mast, round which hay was then piled and set on fire. The revellers threw stones at the pot, and, when it broke, the unfortunate cat had to make its way down through the flames as best it might. It usually escaped with a scorching. *Coitadinho do gatinho, estava todo queimadinho,* as my informant put it with a charming multiplicity of diminutives. Analogies to this custom are not difficult to find, although it is comforting that they are mostly mediaeval rather than modern.

Consideration of the seasonal fires and of their significance can best be deferred to a later chapter. But it should be noted that in many parts of Europe the corn spirit is regarded as being embodied in an animal: wolf, dog, hare, fox, cock, goose, goat, cow, pig, horse, ox, bull or cat. (Round Viana do Castelo a cat used to be burned when all the crops had been threshed.) This may probably be explained by the fact that the custom of killing the god and the belief in his resurrection originated, or at any rate existed, in the hunting and pastoral stage of society, when the slain god was an animal, and survived into the agricultural stage, when the slain god was the corn or a living being representing it. Be that as it may, there is reason to believe that religious processions with animals which were later killed in the character of the god had a great place in the ritual of European peoples in prehistoric times, and that these have left traces in modern folk-custom.

A *boi bento* (sacred ox) figures in several Portuguese religious processions, in that held at Penafiel on the day of Corpus Christi, for instance. At Braga an ox, elaborately adorned, used to be preceded by maidens dancing backwards before him and strewing his path with flowers. He was said to attract more attention than the holy images in the same procession.

At Alter do Chão a festival used to be held on St Mark's Day, which is also observed as the day of *Nossa Senhora do Março* and the first day of Spring, in which an ox was led into church and up to the high altar with the exhortation: *Entra, Marcos, em louvor do Senhor São Marcos* ("Enter, Mark, in honour of Our Lord St Mark"). Here, the animal is clearly identified with St Mark who, in the Balkans and perhaps also in the Iberian Peninsula, has inherited the characteristics of Mars, originally a god not of war but of vegetation. In the rites of Dionysus, among others, the ancients slew an ox as a representative of the spirit of vegetation, and it does not appear unreasonable to suppose that in the origin the *boizinho de S. Marcos* was similarly slain, not as a propitiatory sacrifice but in the character of the spirit of plenty.

Clear traces of animal sacrifice may be seen in the cocks and hens slaughtered in various parts of Portugal at Carnival and other festive seasons. Round Miranda do Douro, the custom of burying hens up to their necks in the earth and stoning them to death has lately been successfully discouraged by the authorities. In many parts of the North, at Carnival time, cocks were suspended from a taut cord and decapitated with swords. To make it more difficult the cock was made to bob up and down, or the executioners were blindfolded. He who succeeded in severing the head was rewarded with the carcase. At the similar *corte dos galos* held at Niza on special Saints' days the swords were hired from the Saint thus honoured. Round Lisbon *cavalhadas* are held at which the mounted competitors tilt at a suspended ring with hens as prizes.

It may not be out of place, at this juncture, to describe two springtide customs connected with the cuckoo. All over the country this bird, perhaps because of its peculiar domestic habits, is consulted by young people who wish to know when they will wed:

Cuco da giesteira,	Cuckoo in the broom thicket,
Quantos anos me dás de solteira?	How long shall I remain a spinster?

At Famalicão (Minho) it is, or was, the custom on March 21st for a peculiar procession to go the round of the village. In a donkey cart went the "cuckoo" in the person of a human representative elevated for the occasion to the court dignity of *Cuco-Mor*. He was followed by a burlesque cortège of masqueraders leading a string of mares laden with scrap iron and domestic utensils. One of these was hung all over with bullock's horns. At every stop the *Cuco-Mor* released caged sparrows and finches, crying, as each flew forth: "There goes a cuckoo for such-and-such a parish." Where there were the most unmarried girls, thither he despatched the largest number of "cuckoos".

At Pragança near Cadaval, a less burlesque festival is held on March 19th, bearing a striking resemblance to the Manx custom of "hunting the wren". A battue is held and a cuckoo trapped, installed in a cart with an old woman on either side, the one spinning and the other winding the wool, and brought in state into the village with a mounted escort of three hundred horsemen.

Connected with the Carnival is another very curious custom. It is practised so near Lisbon as Montijo (the old Alde Galega which so delighted Beckford), where, at mid-Lent, the children arm themselves with pots and pans, and greet with an infernal shindy any old people who are so rash as to venture from their homes on this day. They follow up this manifestation of disrespect, as often as not, with a volley of stones. This pleasant tradition is called *serrar a velha* ("sawing the old woman"). A pamphlet published in Lisbon in 1785 is entitled *Relação curiosa da fugida que faz uma velha para o deserto com temor de ser serrada na presente Quaresma* ("Curious account of the flight of an old woman into the desert for fear of being sawn in this present Lent").

To explain this curious name it is necessary to go farther afield, to the Serra de Monfurado, for instance, where the practice is somewhat different. Here the "rough music" is performed at night outside the houses of old women. The victim is first warned that she is about to die and called upon to make her will and her *acto de contrição* (death-bed repentance). Then one of the party starts to saw a large piece of cork bark which he has brought for the purpose. The saw grates noisily, and the words of the song which accompanies it show that in some way the old woman is regarded as being personified in the wood:

Há que tempos 'stou a serrar	Such a long time I've been sawing
E não vejo serradura;	But there is no sign of sawdust;
Ou a serra não presta	Either the saw's no good
Ou a pele da velha é dura.	Or the old woman's hide is tough.

The same inference is expressed in a different verse sung at Vinhais:

> This old woman of Escaleira
> Is so very smooth and clean
> That when she has been sawn
> She'll make boards for a kneading-trough.

Farther East, one of the serenaders plays the rôle of the old woman, weeping and wailing, making comical bequests to her neighbours ("To the priest, my hide to make a drum"), confessing the most unedifying sins and finally groaning in well-simulated death agony, as the saw cleaves through the bark.

It is difficult to interpret this custom with certainty. The Montijo practice might suggest that it is connected with the custom of driving out Death or the Old Year in one form or another. In one of Gil Vicente's *autos* the Old Year is called *A Velha* ("The Old Woman"). This explanation is supported by the fact that at Tourem (*terras de Barroso*) the straw effigy of an old woman is carried out of the village at the *serração da velha*. The doubling of the real old woman with an imaginary one conceived as immanent in the wood, coupled with a mimic death, suggests, on the other hand, a double presentation of the spirit of vegetation, which in the harvest customs of Europe is often presented as an old woman.[1]

I am more inclined, however, to seek a somewhat different interpretation. Nocturnal serenades of the "rough music" type are in many countries the punishment meted out to conduct of which, for one reason or another, the community disapproves. From Trás-os-Montes to the Alentejo the *chocalhada*, a serenade with prophylactic cow-bells, greets a widowed person who remarries. Moreover, in various regions the Carnival is held to be the most appropriate time for such manifestations of disapproval. The Carnival celebrations at Lagosta (Lastovo), an island in the Adriatic, which include the casting over a cliff of a puppet called Poklad (Carnival), are accompanied by the public recitation of a sort of *chronique scandaleuse* of the year. So, also, in some parts of Beira Alta, is the hanging out of the *compadre*. In the Ardennes the Carnival effigy is often fashioned in the likeness of the husband who is reputed to be least faithful to his wife. At Gouby in Gascony, on Ash Wednesday, a masked figure, clad in rags, who rode through the streets seated back to front on a donkey, used to be taken by force into the houses of those husbands who were reputed to be beaten by their wives. Again, in the French-Basque province of Basse-Navarre, the *tobera-mustrak* or burlesque cavalcades often staged at Carnival time almost

---

[1] In the Minho, at the end of the rye harvest; there used to be held an *Enterro da Velha* (Old Woman's Funeral), in which a straw puppet in skirts was carried round, accompanied by a weeping "widower" and either buried or strung up on a cherry tree.

always centre round some case of adultery, of the marriage of an old or widowed person or the thrashing of a husband by his wife.

The offences which are thus held up to ignominy or ridicule are all, to the primitive way of thinking, contrary to the laws of nature. And it does not seem rash to surmise that in the origin they were punished less on ethical or moral grounds than because of the adverse "sympathetic" effect which they were thought to exert on the crops at the most critical period of their growth. Belief in a secret sympathy between human and vegetable fertility is widespread. In Portugal, for instance, it is thought that if a virgin climbs into a fruit tree, the tree will lose its fruit-fulness. An unfaithful spouse or one pre-viously married and thereby enfeebled, a husband so weak as to let his wife thrash him, such persons are regarded not only as sterile in themselves but as likely to communicate their sterility to the animal and vegetable world. This effect can only  be counteracted by a public expression of disapproval or repu-diation. Sterility is also an attribute of the old, albeit through no fault of their own, and it is significant in this respect that at Vinhais it is only those old women who have no grandchildren who are "sawed". It appears therefore in the last resort that it is their barrenness rather than their age which is thus publicly repudiated, and that the aim is the protection and promotion of fertility.

If, in the Carnival celebrations, a slight preponderance of emphasis is laid on the driving out of Winter and the death of the spirit of vegetation or plenty, those associated with May Day are chiefly concerned with enacting and celebrating the return of Spring and the revival of nature.

Although the May Day festival has no place in the agri-cultural calendar ruled by the solstices, it is as old if not older. There is reason to believe that throughout Europe the solar division of the year was preceded by a simpler terrestrial division into summer and winter. In the pastoral stage of human development, which everywhere preceded the agricultural, the two principal events of the year were the turning out of the flocks to pasture and their return to winter quarters at the beginning and end of summer. It may be surmised that the first of May and November came in this manner to be important dates in the folk-calendar.

Traces of the pastoral origin of the May festival still linger in the remote *terras de Barroso* in the north-western corner of Trás-os-Montes. At Montalegre shepherds decorate the finest yearling ram with flowers and ribbons, and lead him through the streets. At Cortiços the ram has an orange on each horn, and a doll, portrayed in the act of spinning, fastened on his head. Such a custom may have originated in a rite to promote the production of wool. In general, however, it is the cult of vegetable, rather than of animal, fertility which survives to mark the incidence of this originally pastoral festival.

The most obvious symbol of the spirit of vegetation, and one which survives in our own Maypole, is the tree. When the animistic belief that every tree is the body of a separate spirit leads to the polytheistic conception of a general tree spirit which can take up its abode in any particular tree, a stage is reached in which trees, considered as animate beings, are credited with the power of making fruit and crops prosper, herds multiply and women bring forth easily. When, in turn, anthropomorphism displaces polytheism, the same powers are attributed to tree gods conceived in the likeness of mankind.

Both these stages are reflected in Portuguese folk-usage, derived originally from rites in which the tree spirit was conceived as immanent in the tree, as detached from it and represented in human effigy, and finally as embodied in living men and women.

In the Attis cult a pine tree was cut down in the woods, brought into the sanctuary of Cybele and treated as a great divinity. In the same way to-day the peasantry in many parts of Portugal set up a stripped pine trunk on May Day or St John's Day, and dance round it. Attached to the pine tree sacred to Attis was an effigy, the like of which I have seen attached to a *mastro de São João* (St John's mast) at Braganza and to a *galheiro* (a pole decorated the whole way up with rambler rose) at Maia. In this manner the tree spirit is represented in both vegetable and human form, set side by side as though to explain each other.

On the borders of Estremadura and the Alentejo, near Sines, two small dolls are hung from a mast which is set up in June for the *santos populares*. The people call them St Anthony and St John. At Portimão in the Algarve each of two puppets, the one in male and the other in female effigy, is at the top of a separate mast round which the people dance and sing on May Day. These puppets are called *o maio* and *a maia*. Elsewhere a *boneco* has its counterpart in a *boneca* and a *velho* in a *velha*, recalling attempts on the principle of sympathetic magic to quicken the reproductive powers of nature by representing the marriage of sylvan deities. I recall particularly a couple of disguises which I once saw in Lisbon at Carnival time. Two men, one of them dressed as a woman, were carrying "guys", the head and shoulders of which protruded from the front of their coat, while the rump and legs projected from the back. In effect, the effigy appeared to transfix its bearer at an angle and to be, as it were, incorporated with him, so that the spirit of Carnival was represented both in effigy and in human form.

Of human representatives of the May, both male and female, there are, or have been, no lack in Portugal. According to the *Almanach de Lembranças* for 1855 a human *Maio* and *Maia* used to be enthroned by the wayside on May Day in the Minho. At Lagos, on the southern coast of the Algarve, runs a local legend, it used to be the custom of the inhabitants to choose one of their number to be *o Maio*. Round this man's neck they hung all their jewels and golden ornaments, and paraded him through the town with song and dance. This custom was observed annually until, one ill-starred year, a stranger presented himself and offered to play the part, always a wearisome one, of the May. His offer was gratefully accepted by the trusting folk of Lagos. But, alas, the stranger "welshed", taking with him all the valuables of the little town. The custom was abandoned, and for some years the villagers were un-willing so much as to pronounce the name of the month that had brought them such ill-fortune. They preferred to say: " Março, Ábril, *O Mês que vem* ("The coming month"), Junho, Julho...."

At Lagos they told me that this story was the malicious invention of neighbouring towns, but the custom which it describes accords too closely with European usage for this to be easily credible. The *Maio* was clearly an embodiment of the Spirit of Plenty, whose procession round the town was intended as a charm to multiply in every house the riches which he wore about his neck. At Fundão, clothed only in broom branches, the *Maio* used to run through the town like some Jack-in-the-Green,[1] and as *Maio Moco* he is still remembered in many songs, and figured in Tras-os-Montes by a puppet, who has replaced the dancing boy of Alvações do Corgo.

The Alentejo and Algarve were the provinces in which the custom endured longest of electing a *Maia* or May Queen. Adorned with flowers, she was set in a decorated chair at a door or window looking out on to the street, while her satel-lites, with improvised songs, collected money from passers by. In front of the house stood a mast bedecked with flowers and myrtle. Not only each village but each street had its *Maia*, and it was the frequent battles between the supporters of rival queens which led to the prohibition of this custom in the seventies. The *Maia* outlived her legal proscription, for Leite de Vasconcelos saw one at Beja in 1898 and, for all I know, she may still survive in some remote southern village in the *charneca*.

The customs thus far described represent the May as some-thing to be welcomed, almost, indeed, to be worshipped. Side by side with this tradition, however, exists another according to which the *Maio* is an indefinable evil influence, akin to the evil eye and the *ar ruim*, which must be warded off by apotro-paeic means. A common precaution is to hang bunches of broom adorned with ribbons and flowers (which, on the homeo-pathic principle, are themselves called *maios*) on the doors of

---

[1] According to Jaime Lopes Dias he still does so at Tinalhas and Oleiros.

house and stable "to keep the May out".[1] In some parts, where the original purpose of this custom has been forgotten, an aetiological legend explains it on the ground that broom was placed at night upon the house where Christ was sleeping, as a sign to betray Him, but that the following morning every house bore a similar branch. The May can cause a discoloration of the face by entering into the human body, which, however, may be averted by eating sugar almonds or chestnuts before dawn on May morning.[2] The May is sometimes identified with hunger, or with mysterious diseases such as the *quebranto*, and conceived, as in other parts of Europe is the corn spirit, as a wolf, a stag or a donkey. At Santo Tirso, he comes mounted upon an ass, smashes up the pottery and lames the cattle. This dual aspect of the *Maio* puzzled me for some time until a chance clue furnished what seems to be a plausible explanation. In the neighbourhood of Oporto the broom placed upon a house on May Eve is intended to protect it, not against *o Maio* but against witches. Now the broom as a domestic implement, the very symbol of witchcraft, is throughout Portugal made of *giesta* (flowering broom). Thus broom the symbol of witchcraft wards off witches, and bunches of broom called *maios* ward off *o Maio*. The evil aspect of *o Maio* can therefore be equated to the evil power of witchcraft. The puzzling duality now becomes clear. In some parts of the country the outward forms of the agricultural cult which preceded the Christian religion have been preserved unchanged, and their influence is still regarded as beneficent. Elsewhere, Christianity has implanted the belief that they are essentially malefic, and they are feared and warded off accordingly. So narrow is the gulf between worship and abhorrence.

---

[1] At Vermoil broom is replaced by the still more ancient symbol of oak leaves.

[2] To those who seek to trace the origin of "Hear we come gathering Nuts-in-May", I recommend the clear connection discernible in many parts of Portugal between chestnuts and the May Festival. At Trancoso on May 1st the priest used to scatter chestnuts to the children from the church tower. The May boys of Tinalhas and Oleiros are showered with the chestnuts for which their song appeals: *Castanhas ao Maio*. The refrain of the May song of Alvações o Corgo is: *Vamos á caixa das castaninhas* ("Let us go to the chestnut chest"). At Vouzela dried or shelled chestnuts are called *maias*.

CHAPTER VI

SUMMER SAINTS

Most of the ceremonies described in the previous chapter owe their survival to the tolerance of the Church rather than to their incorporation in its calendar. All the long year through, however, an unending succession of religious festivals are held, some of general, some of purely local observance. These *festas* and *romarias* are most numerous during the summer months, from May to September. The *festas* do not differ greatly from the festivities held in other Catholic countries in honour of the Patron Saint of town, village or parish. More characteristic of Portugal are the *romarias*, pilgrimages to particular shrines, often remote from human habitation, which attract the faithful from miles around.

Throughout Portugal, but more especially in the North, these shrines are scattered far and wide. Mostly, they are little whitewashed chapels, standing out like chalk smudges from green hillside or grey rock. Some, such as the Bom Jesus do Monte near Braga and Nossa Senhora dos Remedios at Lamego, have grown rich on pious offerings, and were grandiosely rebuilt in the eighteenth century. Pompous baroque edifices set at the top of long stone stairways, they are approached on the knees by penitent pilgrims and, if only in the spiritual sense, by Mr Sacheverell Sitwell.

Glistening white on the hill-tops, buried in the dim green depths of forests, or keeping their lonely vigil over the gre seas and wind-blown dunes of the Atlantic coast, the little sanctuaries doze the whole year through, until, one summer morning, they are woken to transient life by the gay bands of pilgrims who come to them through the night, beguiling the long leagues with song and dance and the music of guitar, bagpipe and drum. Then for twenty-four hours all is light and noise and colour. When the sun goes down, the earth is illuminated with little cupped lights *à moda do Minho* and in the sky rockets shiver into a thousand stars, until the crowds depart and the shrine is left once more to sink back into slumber.

As in all Roman Catholic countries, these Christian festivals have their roots in pre-Christian cults. In the Iberian Peninsula particular devotion is known to have been paid, in pagan times, to the great Earth Mother, whose ritual, phallic in its essence, took the form of, or itself sprang out of, the worship of stones. To this day Nossa Senhora da Piedade is worshipped as the *Mãe Soberana* (Sovereign Mother) of the Algarve, and the pilgrimage to her shrine at Loulé on August 15th is the most important religious festival of that province. A peculiar feature is a brass band, the members of which run, playing as they go, the whole way up the steep hill on which the chapel is set, a task which, at the height of the Algarvian summer, constitutes a most effective mortification of the flesh.

Everywhere, traces of a cult of stones and rocks survive in the nomenclature of holy images. Senhoras da Rocha, da Penha, do Penedo, da Serra, da Lapa, and so on, abound. These shrines of Our Lady are usually situated on some crag or rock, or near some cave or "rocking" stone, which struck the imagination of the ancient Lusitanians and was worshipped or populated with Mouras Encantadas. For example, Na Sra de Saude at Penha Longa, where a big *romaria* is held in May, stands at the foot of a low spur of the Cintra range, culminating in a sharp pinnacle of rock, and the chapel of the Senhor da Serra at Belas just outside Lisbon is close to the remains of a striking dolmen.[1]

[1] The image of Na Sra do Socorro, venerated in a chapel on the Serra da Enxara do Bispo near Torres Vedras, is said in an old account to have appeared on a rock or in a cavity of that high mountain". In the chapel of Na Sra do

That in other places the worship of Our Lady has superseded the cult paid to the tutelary sprites of springs and wells, is indicated by the fact that at Na Sra da Atalaia in the Ribatejo, Na Sra da Piedade at Gojim near Lamego, and many other shrines, a spring of pure water bubbles up under the very altar of the church.

According to the legend, the statue of the Virgin in the last-named sanctuary was discovered through a dream, in which Our Lady appeared to a local judge on the spot where the chapel now stands. The image was several times removed to a more worthy residence, but miraculously returned to its original site, where a chapel was accordingly built to afford it a resting-place. Images which thus reveal their hiding-place are, of course, a phenomenon familiar throughout Southern Europe, as witness the many *Phaneromene* ikons in Greece. They are no less numerous in Portugal. The legend of Na Sra d'Arrabida is thus recounted, in archaic script and language, on a tablet in her convent church:

The image of Our Lady which is venerated on the High Altar of this church was brought, according to tradition, by an English merchant, who, voyaging from that Kingdom in his ship, carried this prodigious (*sic*) Image, and arriving at this port one night under stress of the weather, all those of the ship beholding themselves in great danger, they sought to have recourse to the said Lady, not finding whom, at the same time there was seen a light on this mountain, and then the weather grew calm and they cast anchor: the next morning, all of them disembarking, they came to the place where they had seen the light, where they found the Lady on a stone; beholding this wonder, the merchant ordered a hermitage to be made on the very spot which we here call "of memory", and he devoted to the service of Our Lady all the rest of his life.

At Merceana, near Torres Vedras, there is a sixteenth-century church dedicated to Na Sra da Piedade. Within, two large panels of seventeenth-century tiles depict the legend of its foundation. A herd of cattle used to pasture on the spot,

Castelo at Aljustrel, there is a large blackish stone on which Our Lady is said to have appeared, and it is thought that the chapel will collapse in ruins if this stone is ever removed. Na Sra do Antime at Fafe (Minho), whose *romaria* is called the Senhora do Sol (Lady of the Sun), is herself nothing more than a shapeless stone rudely fashioned into human semblance.

N. S.ᴿᴬDA PIEDADE DA MERCEANA

Lith-Artistica-Porto
G.J.Machado sculp. Oleyo se Typ Reg Am 1774

and an ox called Mersano was repeatedly seen to fall upon its
knees before an oak tree. One day, the herdsman, gazing at
the tree, saw among the branches a vision of Our Lady with
the dead Christ in her arms. The genuflexions of the ox were
thus explained. Queen Lianor founded a church on the spot,
and until the eighteenth century a special herd of cattle called
the *boizinhos de Nossa Senhora* were kept outside. When Mass
was said the doors were flung open and the cattle driven up,

that they too might partake of it. It is not improbable that this legend was invented to account for some ancient cult of the oak tree and the bull, emblems respectively of the gods of Thunder and War.[1]

The most recent, as well as the most famous, example of such a manifestation is that which occurred in 1917 at Fátima in Estremadura between Leiria and Tomar. On May 13th Our Lady of the Rosary appeared among the branches of a holm-oak to three little children named Lucia, Jacinta and Francisco, and counselled them to prayer and penitence as the best means of combating the misfortunes which had fallen on the world. She promised to reveal herself again on the thirteenth day of every month until October, and did in fact appear accordingly in a cloud of smoke hovering over a spring which gushed forth at the foot of the tree. In less than twenty years Fátima has become the Lourdes of Portugal, to which thousands throng in pilgrimage on the 13th of every month, and especially of May and October, in the hope of witnessing or benefiting by the miraculous cures which are effected by the spring.

In this connection I am reminded of a perfectly true incident which took place in a remote village in the Beiras near where the railway crosses into Spain. One day a small herd girl rushed into the village with the news that while she was minding her flocks an image of Our Lady had fallen down from Heaven beside her. The greatest excitement prevailed, and preparations were at once made to go in procession to the spot, and bring the image with all due honour and reverence to the village church. The priest was at first inclined to be sceptical, but was finally carried away by the general enthusiasm. Guided by the little girl, with the local *Filarmónica* at its head, a long procession went winding over the bleak Beiran uplands, until it reached the appointed place. There, sure enough, was the picture. It was the Queen of Spades, blown out of a train window, no doubt, from the hands of some traveller.

The homing instinct of images such as that of Na Sra da Piedade at Gojim is a common one. At Carregôsa, there is a

[1] An identical explanation is traditionally given for the image of Nuestra Señora de Musquildy near Ochagavia in Spanish Navarre.

S. Bento who returned so frequently to the grotto where he
was discovered, that he was finally given a St John the Baptist,
nominally to keep him company, but in reality to mount guard
over him. The gaoler Saint was not allowed to relinquish his
guard even for the purpose of being carried in his own annual
procession.

The Virgin is worshipped under countless invocations. There
are some, of course, which take their name from a particular
attribute, such as Our Lady of Pity, of Anguish or of Conception. The names of others are derived from the place of
residence, such as Our Lady of Arrábida, Atalaia or Almurtão,
or from a peculiarity of physical geography, such as Our Lady
of the Stone, of the Oak Grove or of the Pine Tree. Others
again have special powers. Na Sra das Febres cures fevers,
and Na Sra do Livramento secures young men exemption from
military service. Our Lady of Childbirth (do Parto), of the
Good Journey (da Boa Viagem), of the Good Death (da Boa
Morte), of the Calm (da Bonança), bestow the gifts indicated
by their names. Other invocations elude interpretation, such
as Na Sra dos Verdes, d'O, do Ar, or do Desterro (exile).
A curious detail, peculiar I believe to Portugal, is that statues
of Our Lord are often named in the same way. Thus Elvas has
its Senhor da Piedade, and Obidos its Senhor da Pedra.

Similar devotion is paid to countless Saints, some of them
from the universal hagiography of the Catholic Church, others
of purely local origin and fame. Among these last one of the
most famous is the *Rainha Santa*, Queen Isabel of Portugal,
whose incorrupt body in its silver casket draws all Beira in
pious pilgrimage to Coimbra in the second week of July.
Another incorrupt Saint, of whom more anon, is S. Gonçalo,
who in the middle of the thirteenth century took up his abode
at Amarante and transformed it from a heap of ruins into a
flourishing city.

The popular beliefs in regard to Saints would alone form
an inexhaustible study, and a fascinating one. The peasantry
are incapable of conceiving their Saints except in their own
likeness, with human virtues and human frailties. S. Pedro
de Rates (the first, and almost legendary, Bishop of Braga)

has a special reputation for revengefulness. On his day not a soul would dare work throughout the Minho, for *o santo é vingativo*, and some misfortune would certainly befall them.

A hundred years ago St Anthony was "still borne upon the staff of the national army, however incredible the absurdity may appear, as a captain in the 2nd or Lagos regiment of infantry".[1] The writer quoted adds that a priestly petition that the Saint might be promoted "in order that the revenue of his chapel might be augmented by the increase in pay" was rejected. Nevertheless, the Saint now holds the rank of colonel of the Cascais regiment.

Like Our Lady, the Saints are generally credited with miraculous powers, and their intervention is sought in all the vicissitudes of life. The realm of this intervention is to some degree partitioned into "spheres of influence", and different Saints are invoked by different classes and communities and to different ends.

S. Antão is the patron of cattle. He is so conscientious that on one occasion, near Taboa, when a priest and a pair of oxen fell over a precipice, he saved the oxen—though not the priest.

S. Fructuoso and S. Romão protect their devotees against mad dogs, S. Gil against shipwreck and snake-bite, S. Dionisio against earthquake, S. Marçal against fires, and S. Lourenço and Sta Barbara against thunderstorms. The latter may be invoked by the following traditional prayer:

> The blessed Saint Barbara
> Dressed and shod herself
> And set forth upon the way.
> She met Jesus Christ,
> And Jesus asked her:
> "Whither goest thou, Barbara?"
> "I go to break up the thunderstorms
> Which wage warfare in the sky,
> To drive them over the Serra do Marão
> Where is neither straw nor grain,
> Nor weeping children, nor crowing cocks."

At Elvas, when it thunders, an old man goes through the streets intoning a chant calling upon the devotees of Sta Barbara

[1] A. P. D. G., *Sketches of Portuguese Life, Manners, etc.*, 1826.

to give her alms. He wears a white surplice, and a tattered
hat. In one hand he carries a palm branch and in the other a
wand on which the likeness of the Saint is painted.

In the same way, the sight is protected by Santa Luzia and
S. Longuinhos, the hearing by S. Ovidio, the skin by S. Bento,
the bladder by S. Vicente, and the throat by S. Blas. S. Amaro
cures lameness, S. Sebastião the plague and S. Paio fever. The
head, together with all mental disease and weakness, is the
special province of St John the Baptist, perhaps because he lost
his own.

As the *advogado* of pigs S. Antão, by an understandable
confusion of names, has usually been displaced by the more
celebrated St Anthony; so that, if you ask a Minho swineherd
if those are his own pigs that he is minding, he will most
probably reply: "Yes, after St Anthony, they're mine."

St Anthony, who was born in Lisbon, has tended, especially
in the South, to displace St John as the *santo casamenteiro*
(match-making saint). In the North both Saints are outvied
by S. Gonçalo of Amarante. True, the latter is primarily a
finder of husbands for old women, as is clear from the song
sung at his *romaria*:

S. Gonçalo d'Amarante,	St Gonçalo of Amarante,
Casamenteiro das velhas,	Marrier-off of old women,
Por que não casas as novas?	Why don't you find husbands for young ones?
Que mal te fizeram elas?	What harm have they done to you?

But when a woman, hopeless of finding a mate by natural
means, has recourse to a thaumaturgical matchmaker, can she
not *ipso facto* be defined as old? Be that as it may, the *romarias*
of S. Gonçalo in January and June are attended largely by
women, who purchase curiously shaped sweets called *testículos
de S. Gonçalo*. The image of the Saint is dressed in monkish
garb, and those who have what the French call *le mal de marier*
tug at the loose end of his girdle, while those who fear sterility
rub themselves against the tomb in which lies his body, incorrupt and pregnant with virtue.

If the Saints are regarded as human beings it naturally
follows that they are treated accordingly. They may be cajoled

and their intervention obtained by prayers and penitence, by vows and votive offerings. But if they fail to accomplish what is desired of them they are exposed to menaces and, in the last resort, to punishment. I am informed that it is by no means unusual, even in good families, for recalcitrant Saints to be turned with their faces to the wall like so many naughty children. The little doves, affixed by a spring to the top of a staff and representing the Holy Spirit, which are kept in every household in the Azores, are, if they fail to accomplish all that is expected of them, penalised by being fastened down with string at Whitsuntide so that they may not jig up and down in time with the dances performed at that festival. Of S. Antonio the matchmaker, an English observer wrote just over a century ago that "if things wear a prosperous aspect his image is honoured with a quantity of tapers; but if the contrary be the case he becomes liable to the grossest possible indignities, and I have even known him plunged into places where his situation must have been anything but pleasant. It is not with lovesick maidens alone that S. Antonio has often to repent of his too extensive reputation; for mariners, who have prayed to him in vain for propitious breezes, at length lose patience and flog his effigy lashed to a mast."[1]

It was only to be anticipated that seafaring people should be less gentle in their methods than landsmen. At Nazaré there 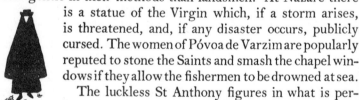 is a statue of the Virgin which, if a storm arises, is threatened, and, if any disaster occurs, publicly cursed. The women of Póvoa de Varzim are popularly reputed to stone the Saints and smash the chapel windows if they allow the fishermen to be drowned at sea.

The luckless St Anthony figures in what is perhaps the most curious of such stories. "The people of Castel Branco", wrote Brydone in his *Tour through Sicily and Malta*, "were so enraged at S. Antonio for allowing the Spaniards to plunder their town, contrary, as they affirmed, to his express agreement with them, that they broke many of his statues to pieces; and one that had been more revered than the rest, they took the head off, and clapped on one of

[1] A. P. D. G., *Sketches of Portuguese Life, Manners, etc.*, 1826.

St Francis in its place, whose name the statue ever afterwards retained."

The *Constituições* of the See of Évora, published in 1534, impose penalties on the practice of carrying images to water in time of drought and threatening them with immersion if they do not bring rain. In some parts of the country this practice still survives, the victims being S. Miguel in the district of Vila Real, S. Antonio in that of Braganza, and Santa Marinha at Segura (Beira Baixa). If the autumn rains come early people say that the makers of straw cloaks have given S. Miguel a ducking, to help their trade. Here it may be surmised that the ducking, though now viewed as a punishment, was originally a rain charm based on the precepts of sympathetic magic. There are indeed places such as Fozcoa, where rocks (which preceded holy images as objects of public worship) are, with the same end in view, turned over or rolled down into rivers. Moreover, at Ligares near Moncorvo it is with no contumely but with all due reverence that a statue of S. Tiago is dipped into the river during a *festa*. At Torreira (Aveiro) a statue of S. Paio is plunged into a barrel of wine, which assuredly is no punishment unless carried to such undue lengths as with the Duke of Clarence of melancholy fame. This last ceremony is performed by women, who afterwards drink the wine and return home singing (perhaps with even more than their wonted *brio*):

Ó S. Paio da Torreira,	O St Paio of Torreira,
Ó Milagroso santinho,	O my miracle-working Saint,
Hei de cá voltar p'r'o ano	I shall return next year
Lavar o Santo no vinho.	To wash you in wine.

Vows and votive offerings for the purpose of enlisting the assistance of the Saints or rendering thanks for their mercies are as picturesque as they are varied. Candles, money and offerings in kind are universal. The last-named are sometimes limited to one particular article. Thus the feast in honour of St Anthony at a village in the Beiras is known as the *festa dos pés de suino* from the fact that only pigs' trotters are offered to the Saint on this occasion. At the *romaria* of S. Bartolomeu do Mar near Esposende, to which small boys are brought and made to bathe in the sea that they may thereby acquire courage, the traditional offering is a black cock. At Santa Marta it consists of flax and at S. Bento of eggs and carnations (a happy blend of the useful and ornamental). Santa Combinha (Macieira) is presented with fleeces by the shepherds whose advocate she is. Finally, to the little white chapel of S. Ovidio on a conical hill which is a landmark from end to end of the Lima valley, pilgrims bring on his day a stolen roof tile.

The virtue thought to reside in stolen articles is a familiar tenet of superstitious belief. At Ponte de Lima it is held that a cow may be helped in calving by eating seven cabbage leaves which must have been stolen and plucked in silence. Goats are thought to prosper on stolen hay; a stolen cat is the best ratter; and stolen plants flourish the best.

The altar of many a little country church or chapel is festooned with ex-votos modelled in wax to represent arms, legs, eyes, breasts, a child's figure or that of some domestic animal not always immediately recognisable. Tresses of women's hair hang at Na Sra da Peninha and Na Sra da Piedade, both near Cintra, and doubtless on many another altar. The walls of chapels and churches which boast a thaumaturgical image are almost always hung with *milagres*, rudely executed pictures depicting and describing, in naïve terms, some escape from death or disaster, attributed to the direct intervention of the Virgin or Saint. One of these pictures, in the convent church of Arrábida, depicts a man walking nonchalantly with a stick in his hand while a soldier in the uniform of the Peninsular War period is firing at him at point-blank range. In the top corner Our Lady is appearing in a glory, and the legend runs: *Milagre q fez N. Snra da Rabida a Luis Naterra do monte Santo, levando Inocente mte hum tiro, por hum Soldado Inglez do Regimento No. 27 o qual a Pegandose Com da Snra ficou Livre do Prigo. No anno de* 1812 ("Miracle wrought by Our Lady of Arrábida on behalf of Luis Naterra of Monte Santo, who being fired at though innocent by an English soldier of the 27th Regiment, placing his faith in the above Lady, escaped from the danger. In the year 1812"). Another shows a cornet player, who, forming part of a band proceeding by boat on a pilgrimage to

Arrábida, fell into the sea and was rescued by fishermen (1865). The cornet is painted with the accuracy of a still-life in a sea of conventional wrinkles. In the same church one may see, with a vividness no less crude, the child who swallowed a pin, the two men who were lost on the mountain (1917), and the man who fell from a cart without injury (1920).

Some of the acts of penance, performed in fulfilment of a vow, take strange forms. A Minho peasant will walk miles to a *romaria* in silence dressed in a winding sheet as though for burial. Another will go even further and pay "four stalwart fellows" to carry him in an open bier.

Self-flagellation and many other extravagances used to be practised in religious processions. Though they have disappeared to-day, it is scarcely fifty years since they were prohibited at Braganza. The author of *Sketches of Portuguese Life* (1826) describes the annual procession through Lisbon of Nosso Senhor dos Passos de Graça and "the groupe of persons who, dressed in horse clothes, crawl on their hands and knees under the images through mud and mire". "I have seen a marchioness and a countess", he continues, "performing this tour de promenade on their bare bones, and flogging or pretending to flog themselves with the disciplines tied round their waist at every pause which the statue made during its progress."

In the procession of Na Sra das Mercês near Cintra, I have seen women, bent almost double, walking beneath the *andores* or stands on which the holy images are carried. At the *romaria* of Na Sra da Póvoa in Beira Baixa many of the pilgrims fell to their knees on arrival, and in this uncomfortable posture made the round of the chapel.

This practice of "circuiting" is one of the many legacies of paganism to Christianity. Connected originally with sun worship, it was intended as a charm rather than as a penance. On August 18th, 1931, I was present at a miniature *romaria* at the village of Janas in the rolling country north of Cintra, half a mile from which stands a little white church already mentioned in chapter III. The church is dedicated to S. Mamede, the local *advogado* of livestock, whose festival day it was. I had heard a persistent rumour to the effect that on this occasion a cock was killed in the church and the assembled cattle sprinkled with its blood. I could find no confirmation of this whatsoever, but peasants assembled from the surrounding countryside, bringing their beasts with them and driving them on arrival three times round the outside of the church.[1] An old *saloio* in a black stocking cap and long white side-whiskers would slowly lead a yoke of plough oxen round the appointed course; a moment later there would follow a string of old women, in red or green velvet jackets and golden chains, riding on donkeys; then would come a flock of sheep, with a mounted drover, bleating and raising the yellow August dust in clouds. The curious thing was that, in common with the Póvoa penitents, they all went widdershins, i.e. left-handed instead of sunwise. In the British Isles such left-hand circuiting is regarded as "sinister", dangerous and unlucky. Nevertheless, one man who started to go round sunwise was sharply called to order. On completing their three circuits, the people went into the church and bought waxen ex-votos in the form of cattle, which they offered up on the altar, and coloured favours with which they adorned their beasts.

A curious feature of the procession of Na Sra das Mercês, previously mentioned, is one in which the notions of penance and votive offering are intermingled. To this *festa*, girls who have so vowed come bearing on their heads *cargas*, which are wooden frames, three or four feet high, not unlike cake stands in shape, and gaily festooned with ribbons, rosettes and artificial flowers. On the different tiers are laid or hung apples and *fogaças*, special wheaten cakes, which are afterwards auctioned for the benefit of the church. The *carga* poised on her head, each girl walks in the procession, accompanied by a male relative in a white surplice, and few of them can return home that night without penitential headaches in expiation of their vows.

Of all the Saints in the Portuguese Calendar, it is, as their name indicates, the *santos populares* who occupy the warmest

[1] At Amendoeira near Macedo de Cavaleiros in Trás-os-Montes cattle are actually driven through a church.

place in the hearts of the people. The three "popular Saints", St Anthony, St John and St Peter, are not only the objects of greatest devotion, but afford the occasion, one had almost said the pretext, for the greatest outburst of revelry. Accordingly, the *mês de São João*, the month of June, on the 13th, 24th and 29th of which these three great festivals occur, is the very peak of the festal year.

It is no coincidence that this month of almost uninterrupted festivities is that in which the days are longest and the sun, reaching its highest zenith, enters into a decline which will continue until Christmas. St John, in Portugal, as elsewhere, is fêted on the very day of midsummer, and the solar character of his feast not only overshadows its religious aspect, but lends many of its features to the festivals of the two other saints, who have thus come to share with St John attributes for which no authority can be found in the New Testament or ecclesiastical history.

It was hardly to be expected that the popular conception of the *santos populares* should bear any relation to the historical and biblical reality. One may seek in vain for a glimpse of the ascetic Baptist or the mystic Franciscan. "The folk", writes Luis d'Oliveira Guimarães, "knows no half measures: either it treats a great man with familiarity and calls him *Tu*, or it does not know who he is." The three saints have been merged in the single conception of a jovial patron of revelry, *um santo muito pândego*, heir to the mantle of Dionysos and Bacchus, with a good taste in wine and an eye for the girls, as depicted in the songs sung at this season:

Santo António com ser santo	St Anthony, though a Saint,
Tambem teve os seus amores;	Loved none the less for that.
Quando os santos namoricam	When even Saints make love,
Que farão os pecadores?	What shall poor sinners do?
São João para ver as moças,	St John, to see the maidens,
Fez uma fonte de prata;	Made a fountain all of silver.
As moças não vão a ela;	The maidens do not go to it;
São João todo se mata.	St John is all aggrieved.
Vá de roda, vá de roda,	Round in a ring, round in a ring,
Haja bailados, cantigas;	Come join in dance and song;
São Pedro, por ser velhinho,	St Peter, despite his age,
Anda ao pé das raparigas.	Is at the feet of the girls.

St Anthony is popularly reputed to have won the maidens' hearts by mending their broken water jars and, if no occasion presented itself, to have done the breaking himself.

It is round St John that these traits have gathered with the greatest clarity and unanimity, and, in the North especially, his is by far the most important festival of the three. In many ways both the feast and the figure of St Anthony are no more than shadows of those of the Prodrome.[1] If St Anthony is a *santo casamenteiro* (marrying Saint), he has filched that attribute from St John. If in the South the sun rises singing on St Anthony's morn, this is only because in the North it dances in the dew on that of St John. It is only in his native Lisbon that the ascendancy of St Anthony is undisputed. And it is in Lisbon, and more particularly in one of the poorer quarters such as that in which he first saw the light, in Alfama or Madragoa, for instance, that I should choose to spend the eve or night of his festival.

[1] If, as I suggest, the festivals of St Anthony and St John are two aspects of the same Midsummer festival, it has occurred to me that the eleven days interval between their dates may be explained by the difference in reckoning between Old Style and New Style which in the eighteenth century amounted to exactly ten days.

For weeks beforehand the children have been making little altars in the street and begging for coppers: *Um tostãozinho para o Santo António!* Such an altar may consist of a wooden stool draped with a white cloth on which are laid all the little ornaments which the children can collect: flowers, vases, coloured postcards, crude religious pictures, and in their midst a clay figure of the Saint, tonsured and clad in his brown. Franciscan robe.

Even at ordinary times the narrow streets clinging precariously to the slopes of St George or Estrela have a theatrical air which is enhanced by the dramatic lighting effects of sunshine and shadow as they fall athwart tall tenement houses. On a festal night their appearance becomes unbelievably fantastic. Festooned with streamers and paper lanterns, the criss-cross maze of alleys is a blaze of light and colour. Every few yards they are bridged by wooden platforms, decorated with green branches and paper streamers, on which, high above the surging throng, brass bands blare out the popular successes of the day. The *arraial* is a sea of faces. Thronging the streets so that one may pass but with difficulty, crowded at their doors and windows, overflowing from the  open wine shops, laughing, shouting, singing, screaming, red and streaming with wine and excitement, the faces mirror in shining reflection the myriad lights. Impelled by some atavistic urge, the people abandon themselves to a dionysiac orgy of revelry. There is something primaeval, violent, explosive in their gaiety. It is the raw material of jollity, unshaped and undirected to a definite end. It suggests to the onlooker a sort of spontaneous combustion of humanity.

While the band plays, the boys and girls of the lower orders dance, close-locked, the dances of to-day. When it ceases they fall back upon the *jogos de roda*, the childish round dances, and the swaying measures of the fisherfolk, *Tirana* or *Verde Real das Canas*. Then, to the tune of a song as blatant and cheeky as Tyl Eulenspiegel's *Strassenlied*, they form lantern processions, forcing their way through the crowds with rough good humour, and leading, as once I saw, a fat, drunken old man, mounted on an ass, who might have been the ancient Silenus or the mediaeval Abbot of Unreason. So, till the dawn, they make merry on the three nights devoted to each Saint, as well as the intervening Sundays, until one wonders how their enthusiasm does not burn itself out.

My first St John's Eve in Portugal was spent at Évora, famous throughout the country for its Midsummer fair. Unfortunately, the fair with its amusements park has proved fatal to the less hardy observances associated with the occasion, and all that I saw was a pathetic little bonfire near the cathedral, over which one lonely old man hopped desultorily, muttering the traditional incantations to the *fogueirinha de São João*.

It was the gypsies who redeemed the evening from bathos. I came upon them among the tents and merry-go-rounds of  the fair. All day long they had been showing off horseflesh to cautious Alentejan farmers, and now at ten o'clock they were dancing to the sound of a steam organ. There were two girls clad in the bright flowery silks beloved of their race, with 'kerchiefs knotted round their heads. One was a beauty of the traditional Spanish type with high cheek bones and golden filigree earrings. The other, with rounded features and grave, slumbrous eyes, had something of the mystery and impassivity of the East. Their partners were dark Romany *chals* who might have stepped straight out of Borrow, but for their dance, a Charleston which would not have disgraced Harlem.

Half an hour later, when I passed that way again, the gypsies had flung off all assumption of modernity and reverted to their ancestral ways. Standing in a narrow circle, they were singing that strange *flamenco* music which is peculiarly their own. Unsmiling now, the girl with the dancing eyes struck a high nasal note with a blow like that of hammer on anvil, held it for a moment against the encouraging *olés* of her companions and then began to descend with little twisting, trilling excursions into ornament, until she fell with a final Phrygian cadence to the foot of the scale. Then the others joined in with a strange wordless chant, the irregular beat marked by the clapping of hands.

The solo part was next taken by a young man, and again there was heard the same wailing, wandering recitative and the same rhythmic chorus with its hint of Eastern drums. Other soloists raised their voices. Sometimes several of them sang in alternation. For a while the singing would appear to become a contest between two of them, while the onlookers took sides as verse succeeded verse, each with a note of arrogant challenge. But always the monotonous reiteration of the refrain lent to their performance a unity and a pattern.

From time to time with an air of finality the group would break up and move away. But it seemed as though the song were stronger than the singers; as though it were inspired by some inward prompting, unconscious and irresistible. Invariably, after a few paces, they would gather round and begin again. At one moment they were joined by a swaggering young giant, whose fortune would be made if Hollywood found him. A long-fringed orange scarf wound round his throat, he pushed his way into the ring and stamped out a few arrogant *flamenco* measures to the singing of his friends. Then from his pocket he drew a bottle of wine and, holding it high above his head, took a long draught. He staggered, with crimson eyes that caught the carbon flares and held them in glittering reflection. Someone snatched his hat from his head, revealing a mass of tousled glistening hair. Another replaced t at a rakish angle. Still he drank until the bottle was nearly empty. Then he dashed it wildly to the ground, and broke away hrough the crowd.

When I returned to my inn, well after midnight, the gypsies vere still singing their Midsummer song. In my nostrils I arried the musky, animal scent of the Romany folk, and in my ars the throbbing beat of their refrain, relentless as the passage f ages.

In the rest of Portugal St John's Eve has retained the character, general throughout Europe, of a Midsummer festival of water, of fire and of lovers. Of these three aspects, it is convenient to begin with the first, for it was its association with water that led the early Church to consecrate to the Baptist this pagan festival. In the Minho, to-day, the division among countless infinitesimal properties of the water made available by irrigation is calculated from midnight on June 24th.

The water which falls from Heaven on this night in the form of dew or rain is credited with every sort of beneficial property until sunrise when its virtue evaporates and vanishes. In order that they may prosper, the children and cattle are washed in it, if they are not actually led down to the river. Used instead of leaven it is thought to make the bread rise. It is "matchless for the complexion" especially to those who have seen the moon reflected in its stilly surface. Accordingly, long before the dawn, women and girls steal forth to *apanhar as orvalhadas* (gather the dew). They wash their faces with dewdrops or at the springs, many of which have been whitewashed anew and decorated with flowers for the occasion. They drink the water "from seven springs" and bring it back home in garlanded jars at the bottom of which she who drew the first water may find a golden ring.

At Briteiros (Minho) folk still think that they can be cured of acne by rolling naked[1] in a field of flax before St John's dawn (near Paredes da Coura this is thought to make a barren woman fruitful) and by rubbing themselves with the dewy stalks. In other parts of the Minho a linen sheet, left out all night to catch the dew, is used for this and other affections of the skin.

Water is also the *raison d'être* of the *cascatas* (cascades) which in the Minho replace the altars of St John. They are usually made in the form of little grottos with moss, flowers, blue paper, bright objects of any kind, mirrors to simulate water and invariably a little image of the Saint.

No less potent than water is the fire lit on St John's night. The Saint's association with fire and light is expressed in his

---

[1] The sacred character of nakedness and its power over evil influences are of course familiar tenets of folk-belief.

dog-Latin title of Santo Johanne Lampadarum, which survives
in the name of a village near Cintra, São João das Lampas. The
Midsummer bonfires are, in Portugal even more than in other
countries, the most indispensable feature of the festival and,
just as the cattle are washed with river or spring water, so the
flocks are driven through the smoke that they may grow fat
and prosper.

At Mondim da Beira the traditional rites are observed
separately by young men and girls. Some days before, the
men go out into the hills and uproot a tall pine tree which,
to the sound of fife and drum, they strip of its branches and
decorate with wild lavender, ferns and other plants. The tree
is then set upright on a hill-top and, on the night of the Saint,
set on fire with pine logs heaped round it. The diversions of
the girls have, in Leite de Vasconcelos' opinion, a decidedly
phallic character. They make a bonfire of dried branches,
plants and herbs, round which they dance, each one crying one
of the following charms as, flinging wide her skirts, she leaps
over the flames:

Fogo no sargaço Saude no meu braço.	Fire in the rock-rose Health to my arm.
Fogo no rosmaninho Saude no meu peitinho.	Fire in the lavender Health to my breast.
Fogo na giesta Saude na minha testa.	Fire in the broom Health to my head.
Fogo na bela luz Saude nas minhas cruzes.	Fire in the *bela luz* (a herb) Health to my thighs.
Em louvor de S. João Que dê saude ao meu coração.	In praise of St John May he give health to my heart.
S. João vai, vem Minha mãe por casar-me tem.	St John comes and goes My mother must marry me soon.

More usually the *facho* or *galheiro*, as the stripped tree is
called, is brought into the village on a decorated ox cart, and
men and girls are mingled in the dances round the fire, over
which the young folk leap, mothers hold out their children,
and sick people extend their diseased limbs, that they may
assimilate the virtue residing therein. This virtue is not ex-
tinguished with the flames, for the dead embers are carefully

collected and used as amulets against storms and for the preparation of magic remedies.

As mentioned in an earlier chapter, puppets or dolls are often affixed to the tree and eventually consumed in the flames. At Paredes da Beira a ceremony called the *queima do pinheiro do gato*, which aroused an indignant protest in the papers in 1932, involves the singeing (at the very least) of a cat in circumstances similar to the barbarous Carnival custom at Arganil mentioned in the previous chapter.

Considered as a feast of lovers, St John's night is chiefly associated with oracles of the "He-loves-me-he-loves-me-not" type. Foremost among these, in the South, are the *alcachofras de São João*, great lavender-blue thistles which are burned, preferably in the bonfires, and set out overnight on balcony or window sill. If by the next morning the charred heart of the flower is found to have put forth new tendrils, you may conclude that your love is reciprocated. I have heard it said that in the natural course of events these flowers blossom twice in quick succession, and that everything depends on whether the flower, when picked, was in its first or second bloom. So prosaic an explanation appears to me unworthy of the Saint in whose name the oracle is worked. In any case there are several alternative forms of divination. It is equally easy, for instance, to pluck a fig leaf and, having passed it three times through the flames of the bonfire, to place it on the roof or in the garden. If the dew of St John falls upon it, this may be taken as a favourable answer. On the same night courting couples are wont to test one another's affections by choosing each a reed and fastening the two together. The one whose reed grows most during the night is the more devoted of the two.

The *sortes de São João* will also reveal to a girl the characteristics or even the name of her future lover. The most usual method is to pour molten wax or break an egg into a glass of water, saying:

São João de Deus amado,	St John of God beloved,
São João de Deus querido,	St John of God adored,
Deparai-me a minha sorte	Reveal to me my fortune
Neste copinho de vidro.	In this little glass.

The shape assumed overnight by the molten wax or the broken yoke will give some indication of the future bridegroom's profession. If it takes the form of a ship, for instance, he will be a Brazilian emigrant; if that of a hammer or a needle, he will be a mason or a tailor; and so forth.

If three beans, one of which is shelled and another half-shelled, are laid under the pillow, and one of them is chosen blindly next morning, the bean with the shell intact will betoken a wealthy husband, that with half a shell a poorer one, and the shelless bean a beggar. In the *terras de Barroso* the girls go round on the night of St Peter and in strict silence knock on the doors of nine different houses. They are then convinced that they will marry the first man whom they see from their window before dawn, or (lest the oracle should be too easily discredited) someone like him. In other places they fill their mouths full of water and, hiding behind a door or standing at a window, with one leg in the air, wait till they hear a man's name spoken, which will be that of their lover.

These forms of divination all presuppose that those who practise them will achieve marriage. Others reveal a less sanguine outlook. A girl can bake three little loaves, hiding in one of them a grain of maize, and place them at random, one under her pillow, one under her door, and one in the street. The next day she will break them open, and according to whether the grain of maize is found to be in the nearest, middle or farthest cake, she will be married soon, late or never. Another method which admits the possibility of a late marriage but not of lifelong celibacy is for a girl to climb up a stairway with a slipper on the point of her toe, which, on arriving at the top, she kicks backwards over her head. As many steps as it falls will years elapse before she marries.

At the Midsummer fairs there are usually to be bought little pots of green marjoram with a paper carnation growing out of each, to which is fastened a scrap of paper containing a love poem in four lines, usually of no greater merit than those found in Christmas crackers. These the people give to one another, and take home "for luck", tending and watering them carefully. Though it may appear far-fetched to connect these with the miniature "gardens of Adonis", charms to promote the growth of agriculture, which were cultivated in the Near East two or three thousand years ago, a link between the two may be found in the pots of grain, on which formerly dolls or Priapus-like figures of paste were placed, which are exchanged between the *comare i compare di San Giovanni* in Sicily and Sardinia. At Catania, indeed, the girls receive pots of basil from their "sweethearts of St John" in exchange for cucumbers, and the thicker the basil grows the more it is prized. When it is recalled that the rites of Adonis were Midsummer rites, a clear connection appears to have been established.

It is evident from the foregoing that there resides in the dark hours of Midsummer night some mysterious virtue which is manifest in all the three aspects of the festival of St John. Though this virtue is held to vanish with the dawn, it may be gathered and retained in certain herbs plucked during the hours of darkness.

The vegetable world plays a considerable part in the folklore of Portugal, as of most other countries. A clear connection is discerned between plant life and the life and welfare of human beings, and is reflected not only in one or two curious locutions, such as *filho das ervas* (child of the herbs) for a foundling child and *ervoeira* for a light o' love, but also in a wide variety of beliefs and practices. It is believed, for instance, that a man who plants a walnut tree will die when the tree attains his girth, and that a scratch from a bramble will only heal when the bush is cut down and burned.

On Palm Sunday in the North, *ramos*, half-hoops adorned with flowers, are blessed by the priests and carried in procession. It is thought that if burnt in stormy weather they will afford protection from lightning. On the same day in the Alentejo, little crosses are set up in the fields with a sprig of rosemary stuck into each of the three points. Ascension Day is popularly known as *quinta-feira da espiga* (ear-of-corn Thursday) from the custom of gathering a posy of corn and olive for plenteous crops, together with poppy and daisy for

peace and silver wealth. On the same day other herbs are
plucked for use in the preparation of simples.

| Todas as ervas têm prestimo | All herbs have power |
| Na manhã de São João. | On St John's morn. |

begins a popular quatrain. In particular, plants with a strong
scent such as rosemary, marjoram, fennel, wild lavender and
elder, gathered on this occasion, are thought to afford protection from witchcraft and thunderbolts, and are used in the
preparation of remedies of magic character. In the North,
branches of holly are cut by people who dance round the tree
singing an incantation which clearly shows the powers with
which the branches are credited:

Azevinho, meu menino,	Holly, my child,
Aqui te venho colher	Here I come to gather you
Para que me dês fortuna	That you may bring me luck
No comprar e no vender	In buying and selling
E em todos os negocios	And in all undertakings
Em que me eu meter.	That I may enter into.

At Travanca, near Feira, there is a tradition that if, on the
night of June 23rd, two men armed with hidden daggers spread
a cloth in the woods, laying a silver coin on each corner, and
retire into a Solomon's Seal which they draw on the ground,
there will come at midnight a host of spirits calling out "Will
you take it up, or shall we take it up?" The men reply that
they will roll up the cloth, and, so doing, make off at full speed.
The next morning, on unrolling it, they will find inside the
seeds of a certain plant which have the gift of making any
woman who touches them incapable of resisting carnal temptation.

Near Guimarães, a woman who wants her hair to grow
longer cuts off the tips on St John's morning and places them
on a bramble shoot. As the bush grows so will the hair grow
by a current of magic sympathy so strong that should the bush
chance to be cut down the hair will wither at the roots.

Quite apart from the analogies which other countries have
to offer, these last two traditions give a very clear idea of the
nature of the virtue resident in Midsummer night. They show

a more or less conscious attempt to harness to human ends, on the lines of sympathetic magic, the powers of nature for

reproduction and growth. A similar intention may be discerned in the cure for a ruptured child, described in a previous chapter, which, it will be recalled, is practised on St John's night. There is no lack of further examples. In Minho and Beira Alta green chestnut branches are planted in the fields on this occasion, in order that the maize may not be blighted or that it may grow to their height. Most striking of all is the conscious attempt, recorded by Leite de Vasconcelos in his *Ensaios Ethnographicos* to transfer the productive forces of nature from a neighbour's field to one's own by climbing on to the *cambão* (yoke support) of a pair of plough oxen and driving them across one's fields, saying:

Aqui vou neste cambão	Here I go riding on this yoke
Na noite de São João	On the night of St John
Para trazer atrás de mim	To draw after me
Pipas de vinho e carros de pão.	Barrels of wine and cartloads of bread.

In order fully to understand the character of the Midsummer festival, however, it is essential to recall the dual aspect which, as we saw in the preceding chapter, the cult of the reproductive forces of nature has come to assume in European folk-lore: on the one hand the widespread survival in a more or less Christian setting of magical practices and beliefs dating from the dawn of human intelligence; on the other a superstitious abhorrence of the forms under which these rites were incorporated in the pre-Christian cults, forms which the Church, well served by an age-old fear of black magic, anathematised under the names of witchcraft and sorcery. The double conception of the "May" as both beneficent and maleficent was a good illustration of this principle. The same phenomenon is to be observed on the night of St John. While, in the Saint's name, the peasants are still performing rites to promote fertility, animal, vegetable and human, they walk in mortal fear of those supernatural beings, phantoms of the human imagination, in whom the pre-Christian nature cult has been personified. Witches, were-wolves, *mouras* and enchanted serpents are abroad in the dark hours of St John, and, as may well be expected, their power is doubled on this night of nights. To ward them off amulets are hung on doors and windows, and the people of Cortiças (Barroso) go so far as to barricade their streets with hurdles, doors, ploughs and similar objects.

This duality in the ceremonies of St John's night has led to divergencies of opinion regarding their exact interpretation. The use of fire and water, in particular, admits of several possible explanations. The bonfires may on the one hand be regarded as magical rites intended to preserve the sun from the decline into which he enters after the day of his greatest splendour. It was Mannhardt's theory that the aim of the festival was to provide by sympathetic magic a plentiful supply of sunshine. The theory has this in its favour that it accounts for the occasion of the festival, but it scarcely appears to furnish an adequate explanation of its character. Frazer points to the fertilising influence exerted by the fires. "Whether applied in the form of bonfires blazing at fixed points, or of torches carried about from place to place, or of embers and ashes taken from the smouldering heap of fuel, the fire is believed to promote the growth of the crops and the welfare of man and beast, either positively by stimulating them, or negatively by averting the dangers and calamities which threaten them from such causes as thunder and lightning, conflagration, blight, mildew, vermin, sterility, disease and, not least of all, witch-craft."[1] But between the "positive" and "negative" poles he halts, undecided, and there is, indeed, considerable room for doubt.

On the one hand, the fires appear to have a direct fertilising influence, due perhaps to their having been originally the means of the destruction meted out to that god (or his repre-sentative) whose death and resurrection were regarded as

[1] *The Golden Bough*, Abridged ed., p. 642.

essential to human welfare, and who survives in Portugal in theriomorphic form in the cat of Paredes da Beira, and in anthropomorphic form in the puppets burned from Braganza to Vila do Bispo. "When the God happens to be a deity of vegetation there are special reasons why he should die by fire. For light and heat are necessary to vegetable growth, and, on the principle of sympathetic magic, by subjecting the personal representative of vegetation to their influence you secure a supply of these necessaries for trees and crops." At this point the author of the *Golden Bough*, who has hitherto been in general agreement with Mannhardt, hesitates, impressed by the "negative" theory put forward by Westermarck, that the fires, while intended to promote the growth of vegetation, do so by cleansing it of all baneful influences. This theory has the support, of doubtful value though it is, of many who themselves light the Midsummer fires to-day.

It is to be noted, moreover, that the evils against which the fires, considered as negative agents, afford protection are all prerogatives of witchcraft. Now witchcraft is only a metamorphosis of that cult of fertility which the fires, considered as "positive" agents, are held to promote. It does not seem, therefore, to matter very greatly whether they are to be regarded as a stimulant or as a disinfectant. They may well be both.

The same may be said of the Midsummer water, round which the battle has raged less fiercely. Even when it lacks the special virtues with which St John endues it, water is widely regarded as a fertilising element, as has been shown in a previous chapter. But, like fire, it is also a cleansing element. It is surely enough that the three aspects of the festival of fire, water and lovers, have in common the purpose of promoting fertility, both indirectly by furnishing a supply of food from the vegetable and animal world, and directly by assuring the reproduction of the human race.

Before taking leave of St John, I cannot resist the temptation to quote a few of the verses sung round his bonfires in all parts of the country. They reflect not only the traditions associated with the cult of the Baptist, but also its intimate, tender, at times almost mystical nature. The Saint is portrayed as participating with childish delight in his own festival, and such is the affection of the Portuguese maidens for their *santo casamenteiro* that they come very near to making love to him:

São João adormeceu	St John fell asleep
No regaço de Maria.	In Mary's lap.
"Acorda, João, acorda,	"Awake, John, awake;
Que já lá vai o teu dia."	Here dawns your day."
Donde vindes, ó São João,	Whence come you, O St John,
Que vindes tão molhadinho?	Whence come you all wet with dew?
Venho de ver as fogueiras	I come from watching the bonfires
E colher o rosmaninho.	And plucking wild lavender.
Para fazer as fogueiras	To build the bonfires
Na noite da sua festa	On the night of his feast
São João traz lá do monte	St John brings down from the mountains
Um braçado de giesta.	An armful of broom.
Até os Mouros na Mourama	Even the Moors in Barbary
E os Turcos na Turquia	And the Turks in Turkey
Festejam o São João	Celebrate St John
Como nos cá no seu dia.	As we do here on his day.
Na noite de São João,	On the night of St John,
É que é tomar amores,	This is the time to make love,
Que estão os trigos nos campos	For the very corn in the fields
Todos com as suas flores.	Is all in flower.
O São João de Figueira	The St John of Figueira
Não tem velas no altar;	Has no candles on his altar;
Se o santo me casar cedo	If the Saint marries me soon
Sou eu que lh'as vou levar.	'Tis I who will take them to him.
São João me prometeu	St John made promise to me
De me dar um bom marido,	To give me a good husband,
Vou-lhe lembrar a promessa	I will remind him of his promise
Pois o santo é esquecido.	For the Saint is forgetful.
Casai, rapazes, casai,	Come marry, my lads, come marry,
Que as noivas baratas são;	For brides are cheap;
Cada três por um vintêm	Three of them for a penny
Na noite de São João.	On the night of St John.
São, João, São João, São João,	St John, St John, St John,
Não deixeis este verão passar,	Don't let this summer pass,
Dai-me noivo, São João, dai-me noivo,	Give me a sweetheart, St John, a sweetheart,
Dai-me noivo, que eu quero casar.	A sweetheart, for I would wed.

Ó meu São João da Ponte
Ó meu santo pequenino,
Heis de ser o meu compadre
Do meu primeiro menino.

O my St John of the Bridge
O my little Saint,
You shall be the godfather
Of my first child.

São João adormeceu
No regaço de sua Mãe,
Quem me dera São João
Dormir contigo tambem.

St John fell asleep
In his mother's lap,
Oh, that I too, St John,
Might sleep with you as well.

Abaixai-vos, carvalheiras,
Com a rama para o chão:
Deixai pasar as romeiras
Que vão para o São João.

Bend down you woods of oak,
Sweep your branches on the ground:
Give passage to the pilgrim maids
Who are bound for St John.

CHAPTER VII

THE DEATH OF THE YEAR

The last quatrain in the preceding chapter brings us back to
the *romarias* of the North, most numerous in July and August,
which we may best approach in company with the gay groups
of peasants who dance their way across the countryside, singing
verses appropriate to the occasion:

Ó Senhora da Lapinha
Que dais aos vossos romeiros?
Dou água das minhas fontes
Sombra dos meus castanheiros.

O Lady of Lapinha
What is your gift to your pilgrims?
I grant them water from my springs
And shade from my chestnut trees.

A Senhora da Abadia
Tem o seu pilar de pedra;
Bem o pudera ter de oiro
Ou de prata se quisera.

Our Lady of Abadia
Has a pillar of stone;
She might have one of gold
Or of silver if she wished.

Senhora Santa Combinha
Que lá mora no altinho;
Por maior que seja a calma
Sempre lá corre ventinho.

O Lady Santa Combinha
You dwell up there on the height;
However great be the stillness
There blows there a little breeze.

Senhora Santa Combinha
Tem um manteuzinho branco
Que lh'o deram as dueiras
Domingo do Espirito Santo.

The Lady Santa Combinha
Has a little white cloak
Which the herd-girls offered to her
On the day of Pentecost.

Senhora da Piedade,
O caminho pedras tem:
Se não fossem os milagres
Já cá não vinha ninguem.

Our Lady of Pity
Her path is strewn with rocks:
Were it not for her miracles
No one would come hither.

Nossa Senhora da Póvoa	Our Lady of Povoa
Bem me podeis perdoar:	I crave your pardon:
Vim a vossa romaria	I came to your *romaria*
Só p'ra cantar e bailar.	Only to sing and to dance.
Senhora Santa Combinha,	O Lady Santa Combinha,
De lá venho eu agora,	Thence come I now
Em manguinhas de camisa	In my shirt sleeves
Tocando numa viola.	And playing a guitar,
Senhora do Almurtão,	Our Lady of Almurtão,
Nossa Senhora raiana	Our Lady of the frontier
Voltai as costas a Castilha,	Turn your back on Castile,
Não queirais ser castelhana.	Do not be Castilian.
Nossa Senhora do Carmo	Our Lady of Carmo
Aonde a foram pôr?	Whither have they taken her?
Entre Arelvas e Ourondo	Between Arelvas and Ourondo
Onde não há outra flor.	Where blows no other flower.
Virgem Senhora da Póvoa	Lady Virgin of Povoa
Mandai o tempo alegre;	Send the weather fine;
Há oito dias que chouve	For eight days it has been raining
Água fria como neve.	Water as cold as snow.

Who shall say when or how these pilgrimages began? The *romaria* of Na Sra da Guia, at the mouth of the Ave, has been held continuously since the eleventh century. They are frequently mentioned in the lyrical verse of the troubadour period. A thirteenth-century *bailada* (dance-song) ascribed to Pero Vivianez conjures up their arcadian charm as freshly as though it had been written to-day:

Pois nossas madres van a San Simon	Now that our mothers go to San Simon
De Val de Prados candeas queimar,	Of Val de Prados, candles there to burn,
Nós as meninas punhamos d'andar	We maidens too set forth upon our way
Con nossas madres, e elas enton	Bearing our mothers company that they
Queimen candeas por nós e por si,	May burn their candles for themselves and us,
E nós meninas bailaremos i.	And we the maidens there shall tread the dance.

A *romaria* falls into two distinct sections, the purely religious celebrations, and the secular *arraial*, a name given to the fair in the precincts of the church and to the revelry held there on the eve of the festival. These saintly vigils have provoked scandals at intervals ever since the re-conquest and, round Braga, in its time a great Roman settlement, were known till well on in the Middle Ages by the significant name of *Kalendas*.

In his *Historia da Administração Pública em Portugal nos seculos xii a xv*, Gama Barros states that, throughout this robust age, the folk "ate and drank in the churches, setting up tables, and sang and danced there on the pretext of the vigils of Saints, or on the day of some festival: and the very parish priests kept corn, barley, wine, olives, chickpeas, onions, garlic and other similar products in the churches for more than a day".

The *romarias* have many points in common with the seasonal expulsions of evils which were so prominent a feature in the religion of Ancient Rome, as indeed in all primitive cults. That attendance at a *romaria* acts as a sort of ceremonial cleansing is amusingly indicated by a popular quatrain:

Eu venho da romaria	I come from the pilgrimage
Da Senhora da Canhota;	Of Our Lady of the Tree-trunk;
Agora qu'eu venho santo	Since I come thence sanctified,
Dá-me um abraço, cachopa.	Give me a kiss, my girl.

Fire and noise are the means most usually employed to drive out evils, and there is plenty of both at *romarias*, either singly or combined in rockets and fireworks, which latter the Portuguese have raised almost to the level of a fine art.

Of undiluted noise Pinho Leal quotes a good example in the *Tim-tiri-nó* at Capinha, where, all through Advent, the peasants observed the custom, of which no priest could cure them, of keeping the church bells jangling without a moment's intermission by day or night.

The deliberate cultivation of noise must account originally for the brightly painted clay whistles, made in and near Barcelos, which are on sale at every festival or fair throughout the country from Whitsuntide till October. In form some of these whistles are astonishingly archaic. A bird with little ones linked to it by earthenware loops reproduces almost exactly a Graeco-Roman toy in the National Museum at Athens. A horse which might have been copied from that of Troy, a

bull of Minoan aspect and many other figures may be compared
not only with the modern clay figures made at Seville, in the
Balearics and in Mexico, but also with ancient ones from
Cyprus, Crete and the Crimea.

Another of the essential features of the *romaria*, the *merenda*
or picnic meal eaten in the open, links it with pagan festivals

at which offerings of food were eaten in common by all the
participants, who thereby hoped not only to accomplish an act
of propitiation but also to acquire strength against evil influences.

The most remarkable Portuguese example of a votive banquet was the *bodo do Espirito Santo* at Santiago de Cacem on
Whit Sunday, which is mentioned in documents as far back as
1404 and was last held towards the end of the nineteenth
century. This was a feast offered to the poor and prisoners after
the dishes of which it was composed had been carried in
procession round the town. The sacrificial nature of the proceedings is clear from the following extract from an eyewitness's account written in 1862: "In the place where the cows
are slaughtered to provide beef for the feast all the diseased
collect, armed with pots, pans and dishes which they fill with
the cows' blood. As soon as they obtain this, they anoint with
it not only the affected parts but also those parts which are no
affected in order that these may be smitten by no disease . .
so that it is not unusual to see men, women and children with
their faces a crimson mask. . . ."[1]

[1] O Bejense, No. 81, of July 12th, 1862, quoted in *Revista Lusitana*, XI, pp. 71–3

Such orgies are happily obsolete. But there is often a decidedly pagan quality in the *romarias*. "One does not find", writes Luis Chaves, "a single Christian festival without some pagan aspect or flavour".

An orgy of a less offensive sort is the burlesque festival of Santa Bebiana, still held in 1932 at Lousã. Santa Bebiana is said to have been the sister of S. Martinho, on whose day the festival falls. It is not in the Calendar of Saints that she should be sought, however, but in a dictionary among the derivatives of the verb *beber*, "to drink". S. Martinho himself is something of a patron of revelry, since it is he who holds sway over the annual slaughter of the pig which is always made the occasion for a festive repast. Santa Bebiana presides more specifically over the potations. At Lousã the lads of the village formed a torchlight procession, with one of their number, in female disguise, hoisted on their shoulders in a wheelbarrow. In her hand this unedifying Saint held a flask which she applied to her lips at frequent intervals. This parody of a religious procession concluded with a mock sermon and a hymn:

Todo o povo bebe vinho	The whole town drinks wine
Ao domingo e á semana	On Sunday and all through the week
Em louvor de São Martinho	In honour of St Martin
E de Santa Bebiana.	And of St Bebiana.

Of another *romaria*, in no way burlesque, one may read that "the diversions, the feasting, the too openly expressed gallantry during the whole festival, 'with consequences at the end of nine months', as they say in Terroso, the quarrels, feuds and disorders, have brought about the prohibition of this positively orgiastic festival"[1].

Too often, indeed, as in Gil Vicente's poem:

El mozo y la moza van en romaria,	The lad and the maiden go in pilgrimage,
Tomales la noche naquella montina.[2]	And the night overtakes them on the hillside.

It was the pagan licentiousness of the nocturnal *arraial* which led to an edict by the Archbishop of Braga in 1932,

---

[1] In *Portugalía*, ii.   [2] Gil Vicente, *Triunfo do Inverno*.

prohibiting all religious celebrations and even the opening of the churches at any festival on the eve of which there had been an *arraial*.

It is not for a foreigner to question the wisdom or propriety of this decision. It is only too obvious that a night of dancing and feasting with plentiful libations of *vinho verde* is anything but an ideal preparation for the Mass and procession of the morrow. On the other hand, religion has generally found it expedient to throw its weight on the side of conservatism, since innovations are rarely to its advantage, especially in an age of unbelief. It is difficult for the peasant to see the *romaria* as anything but a complete whole, and if the edict is enforced more strictly than similar prohibitions have been in the past, one can only fear that it will mean the disappearance of one of the most characteristic and colourful Portuguese institutions, and one which, for all its abuses, is a triumphant expression of simple, unquestioning faith. "The folk cannot grasp so much spirituality", writes a Catholic priest, the Abade Baçal; "and ceremonies and enthusiasm without revelry are out of the question: there will soon be nothing left...."[1] It argues great courage on the part of the Church to face this risk.

A type of religious festival peculiar to Estremadura, which has lost much during the last twenty years, is the *cirio*, the origin of which is explained in an eighteenth-century manuscript in the National Library:

The discovery of numerous images of the Holy Virgin, hidden for many centuries after the invasion of the barbarians, awoke the devotion of the people, who built churches and instituted new pilgrimages. And since there was competition between different places and parishes, each place had in the church a great taper, which was lit on the occasion of the festival. This taper was called a *Cirio*, a name which passed to the corporation of the pilgrims who on a particular occasion went to do honour to Our Lady.[2]

The most important are the so-called "royal" *cirios* of Na Sra de Nazaré and Na Sra do Cabo d'Espichel. No less than twenty-eight parishes send *cirios* to the latter shrine, and those

[1] Padre Francisco Manuel Alves, *Trás-os-Montes*, p. 25.
[2] Quoted in Luis Chaves, *Portugal Além*, p. 126.

Plate IX

MOURISCOS AT SOBRADO

THE MOORS IN *FLORIPES*

Plate X

THE DANCE OF KING DAVID AT BRAGA

THE *GOTA* AT CARREÇO

which visit Nazaré at intervals from August 5th to September 15th come from places as far apart as Lisbon and Coimbra. Each of these vassal parishes has the honour of housing the sacred image for a year three or four times in a century.

In the old days the *cirios* must have been an imposing sight as they went in peregrination through the countryside. Early this century they were headed by mounted standard-bearers in frock coats and top hats. Then in a carriage came children dressed as angels who at every stop sang *loas*, traditional hymns in honour of the Virgin. Various *cirios* had their own

peculiarities. One carried the rich church plate of Mafra all the way to Nazaré, and on arrival drove three times round the church and distributed food, wine and money to all present.

The advent of the motor-car has robbed of all colour the *Cirio da Prata Grande* and most of the others. The *anjinhos* now consist of a lorry-load of children who do not even dress for the part, much less act it. The only festival in Estremadura at which the *cirios* still retain something of their traditional character is that of Na Sra de Atalaia held on August 29th. Those from Lisbon are ferried across the Tagus in graceful sailing barges. The *Cirio da Quinta do Anjo* trundles slowly across country in a gaily festooned ox cart with an escort in old costumes, while that of Carregueira is brought by young men on horseback with girls riding pillion in Andalusian fashion. The *caloiros*, those who are making the pilgrimage for the first time, are swung by the arms and legs and bumped none too gently against a certain pine tree near the sanctuary.

Not unnaturally, the strictly religious part of all these festivals is from the ethnographical point of view less interesting than the secular. There is little or no difference in the celebration of a Mass between Trás-os-Montes and the Algarve, beyond the degree of religious fervour which diminishes progressively from North to South.

The procession, on the other hand, goes back far beyond the origins of Christianity. Its original intention may have been magical (on the principle of circuiting) rather than propitiatory or devotional. Traces of this intention may still be discerned to-day, when the transportation of holy images from one chapel to another is regarded as exercising a direct influence on the elements or the crops.

In 1932, after an unprecedented period of torrential rain, the inhabitants of Penamacôr to the number of two thousand accompanied the image of Na Sra do Inemex from her chapel to the parish church. From the day (the first Sunday after St Anthony) when Na Sra da Lapinha is brought from her chapel into Guimarães, the blight is thought not to attack the maize.

The protection of the crops by circuiting is the aim of rogation processions in many countries. These are known in Portugal as *cercos, clamores, ladainhas* and *caramões*. Banners, crosses and holy images are carried round the fields with prayers and litanies. At night the participants carry torches, and in bygone days there was much firing of muskets. Such processions may be held at set times or seasons or during unwonted tribulations, the peasants hoping, in the words of an eighteenth-century Portuguese writer, "thus to be free from baneful insects and intemperance of the air, damaging their crops, killing flocks and afflicting the people". At the *clamor* held at Refojos, near Ponte de Lima, on Easter Thursday, the peasants shout, "St John the Baptist protect our grain".

More usually, the procession has no such specific purpose and is purely an act of worship. There is no great scope for variety or local colour in the painted images, the brotherhoods with scarlet or purple surplices who carry them, or the little children dressed as angels who follow in their train, in some cases almost as soon as they can walk. Such picturesque details as are retained in exceptional cases are derived from the pageantry and extravagance of the Middle Ages. For instance, in an annual procession held at a village near Pedras Salgadas in Trás-os-Montes, two of the participants represent Adam and Eve, and are dressed (more or less) accordingly.[1] At that of Na Sra de Parada near Montalegre, the Virgin rides on an ass as though to Egypt and is followed by a yellow-clad man bound in chains, representing the "Philistine".

Very strange processions used to be held at Travessó near Aveiro for the festival of the Holy Martyrs of Morocco (d. 1220). A procession of male penitents, naked to the waist, was prohibited by the bishop last century, but a similar procession of women, their stripped torsos "honestly covered with white linen towels", still took place in Pinho Leal's time. An eyewitness who saw it perhaps twenty years ago relates that the women went in single file, at night or in the early hours of the morning, each carrying a tile with which she prodded the one in front in a curiously suggestive manner. There was also a procession of "Moors" with a *Rei Mouro*, dressed in yellow and wearing a huge head-dress of feathers.

In Portugal the richest and most elaborate of the mediaeval church processions was that of Corpus Christi. In 1844, according to Antonio Feliciano Castilho, it still retained at Vila Cova "a large number of its extravagant accessories of three or four centuries ago; a *Serpe*, dancing Moors, etc., etc." The *Serpe* or Dragon is still preserved in the Corpus Christi procession at Penafiel, which also includes cardboard giants and traditional dances by professional Guilds such as the *dança dos alfaiates* (tailors' dance) and *dança dos ferreiros* (smiths' dance). On the same date a fearsome pantomime dragon, operated by men concealed within its pasteboard flanks, appears at Monção on the River Minho, where it is fought and vanquished by St George. It is called the *Santa Coca*, but belies

---

[1] Eve wore a white cotton dress with fig leaves sewn on it.

its sanctity by uttering all sorts of obscenities for the delectation of the crowd.

The responsibility for furnishing the various features of the Corpus Christi procession devolved in the Middle Ages upon the Guilds, which came in this way to form a link between the pagan mummery of the past and the stylised survivals of to-day.[1]

Many of these features appear in the Middle Ages to have been transferred to the procession held on St John's Day. The procession of St John at Braga in 1579 included, besides the usual candles and decorated carts, a *Dança das Pelas*, a *Mourisca*, a *Serpe* and *cavalinhos fuscos* (hobby-horses). These last are familiar figures in the ritual dance of Europe. They have practically disappeared from Portugal to-day, and I have seen only rather sophisticated examples in the poorer parts of Lisbon at Carnival time. But in the Middle Ages they figured in many processions. The *dança das pelas* consisted of girls dancing on the shoulders of men, like those who figured in the Setúbal Corpus Christi procession in 1484, "surrounded by others running about like bacchantes, leaping and playing tambourines and drums", or for that matter like the dancers of the *Moxiganga* near Tarragona to-day.

The *Mourisca* consisted of a group of young men with their "king", armed with shields and lances, who acted a mimic battle. They were in Portugal, as in Provence, a traditional feature of the Corpus Christi procession. At Braga in 1532 they were reorganised into a regiment of twenty at the expense of the town, furnished with uniforms to replace the rags in which they had been wont to appear, and ordered to dance in several religious processions during the year.

To-day all these picturesque elements have disappeared from the Braga procession, which nevertheless retains two interesting features, the *carro das ervas* or *dos pastores* and the *dança do Rei David*. The procession is intimately linked with

[1] Thus at Coimbra in 1517 the following Guilds each furnished an item in the procession: carpenters, potters, tailors, weavers and shoemakers. At Oporto in 1621 pastrycooks, tanners, smiths, masons, hatters, vintners, merchants and shopkeepers are also mentioned. (Sousa Viterbo, *Artes e Artistas em Portugal*, pp. 244-5.)

the sacred dance, which in its turn was a stylised form of ritual pantomime. It is scarcely surprising, therefore, that both dance and mime should have their place in the older processions,[1] and it is in this light that both features of the Braga procession should be viewed. In 1932 it was suppressed by the archbishop, but fortunately the two most interesting features were suffered for tradition's sake to go round the streets, shorn of none of their character though deprived of the setting which gives them their real *raison d'être.*

The *carro dos pastores* proved to be an elaborately decorated lorry, drawn by oxen, on which was erected a large and unsubstantial stage rock, with a fountain in a niche. Half way up stood a small boy representing St John, dressed in silks and velvets, with periwig and painted face, leading a live lamb adorned with red ribbon. At intervals this very rococo Baptist sang in a forced voice a conventional hymn, while a group of children dressed as shepherds and shepherdesses danced a simple country dance. At this the top of the rock fell miraculously apart. A child angel shot up unsteadily, his feet resting insecurely on a white woolly cloud, on which, for a moment, he stood poised in an attitude of benediction. The *pastores* then processed round the rock, and this pocket "Mystery" came to an end.

King David and his ten courtiers wore a costume resembling the mediaeval artist's idea of Eastern dress. Their coats were of blue or crimson velvet, trimmed with silver braid, and except for the king, who bore a crown, they wore turbans wound round a kind of fez like the magi in a Renaissance Adoration. They each carried a musical instrument, violin, guitar, flute, triangle or cello, on which they played as they marched through the streets. Every now and then they stopped, formed up in two lines, and, continuing to provide their own music, began to dance. Thrumming a guitar, King David gravely hopped his way down the line. Then with a back-step he retired, concluding with a pirouette which sent his ermine-trimmed

---

[1] The more sophisticated processions held in Lisbon on civic occasions continued till the eighteenth century to incorporate elaborate allegorical dances such as those of the Rivers, Cities, Seasons, Muses and others.

cloak swirling round him, as the decorous little tune came to an end.

Two of the courtiers now repeated the figure, first by themselves, and then with the king, while the others "cast off", wound round and filed back to their original places. Thus concluded a survival as remarkable as the Dance of Seises in the cathedral of Seville.

In spite of its unmistakable air of antiquity, the performance is not mentioned by name before 1726, although there is a reference dating from 1632 to a dance the performers of which played musical instruments, and King David figured in the Corpus Christi procession at Oporto in 1621.[1] Perhaps it is a modern development of the *Mourisca*. The rôle of King David is hereditary in a family living not in Braga itself but in one of the neighbouring villages.

It is easy to see how the Biblical allusion to David dancing before the Ark could have been used to explain and justify a pre-existent ritual which it was found expedient to retain. A similar but more rustic dance which is not connected with King David but is called the *Dança do Genebrés* is performed annually at the *romaria* of Na Sra dos Altos Ceus at Lousa (Beira Baixa). The dancers carry instruments, but wear the characteristic tall headgear of the "mauresque" dancer. At the conclusion of their dance the leader unsheathes a sword.

Dances by the priests of Amarante in honour of S. Gonçalo were discontinued long ago, and of the few religious dances performed within living memory in various parts of the country the greater part failed to survive the turn of the century.

At Arcozelo da Serra, near Gouveia, the procession held on the feast of the Assumption of the Virgin (August 15th) included no less than four different groups of dancers called *charolas*, which performed at each stopping-place. The *dança das donzelas* consisted of six or eight little girls who pretended to be Moorish maidens desirous of embracing the Christian faith, and enacted a little comedy in which they were baptised and sprinkled with holy water by a small boy dressed as an angel. The *dança dos marujos* was performed by boys repre-

[1] Sousa Viterbo, *Artes e Artistas em Portugal*, p. 245.

senting sailors who had escaped from shipwreck thanks to the intervention of Our Lady and were thus testifying their gratitude. The performers of the *dança dos pretos* were men with blackened faces, dressed in red and hung with little bells. To the strains of the guitar they marched, danced the fandango and indulged in all sorts of antics. According to Pinho Leal, their rôle was that of slaves complaining to Our Lady of ill-treatment at the hands of their masters, and their "farce" was interlarded with obscenities uttered even in the presence of the Host. The *dança dos espingardeiros* consisted of eight or sixteen young men who acted a battle between Portuguese and Spaniards, in which the former, invariably victorious, magnanimously spared the lives of their enemies.

Even more than these, one must regret the discontinuance of the *Mourisca* which used to be performed on St John's Day at Pedrogão Pequeno, where seven men entered the church, exotically dressed in silk, with ribbons, and open jacket, diagonal shoulder-straps and a conical hat decorated with flowers. Two of them played guitars, two tambourines, and two carried staffs surmounted by a bunch of carnations. The seventh was called the *Rei da Mourisca*, and wore a sword, a shield, a royal mantle and a crown. Bowing low before the statue of the Saint, they executed a slow and decorous dance, at the conclusion of which they made a genuflexion before the altar, and the "king", pirouetting on his left foot, shouted, *Viva meu compadre São João Baptista!*

The constant recurrence in mediaeval festivities of "mauresque" dancers who were Moorish only in name is a widespread phenomenon for which no satisfactory explanation has yet been found, though it may perhaps be linked with the interpretation already given of the *Mouras Encantadas* (see p. 80). Isolated groups of these dancers still remain scattered across Southern Europe, from Portugal through Spain, Southern France and Italy to Dalmatia, where an annual *Moreška* is performed at Korčula (Curzola). If the derivation of "Morris" from "mauresque" may be accepted, they survive also in those parts of England where the Morris dance still retains its primitive, authentic robustness. Wherever they still appear in the flesh, and wherever, in the yellowing documents of the past, they are shown to have existed, "mauresque" dancers have several points in common. They are invariably young men, performing, on special occasions only, a dance which is serious in purpose, ceremonial and spectacular rather than recreational in character, and from participation in which women are strictly excluded. Almost invariably they wear a special costume comprising some of the features proper to ritual dancers: gaily coloured ribbons and artificial flowers, tall headgear more or less like crowns, prophylactic mirrors and bells, and often enough swords, or the little sticks to which, by a process of atrophy, these last have so often shrunk.

The sword dance, which in a number of cases concludes with the simulated death of one of the dancers, is generally regarded as being a survival in a conventionalised form of a fertility rite originally attended by human or animal sacrifice. Professor Elliot Smith, indeed, has put forward the theory that the whole ritual of dancing was originally suggested by rites enacting the death and resurrection of the corn-god Osiris. Of late, however, Dr Richard Wolfram has suggested that the sword dance derives from the men's secret societies with their initiation ceremonies. In either event the origins of the sword dance, as performed in almost every European country, are to be sought in the most primitive pagan ritual.[1]

During the Carnival of 1933 I saw two dances which were clearly degenerate forms of the *Mourisca*. The *dança da Luta*, revived in Lisbon on this occasion for the first time since before the war, showed unmistakable signs of having been remodelled, probably in the eighteenth century, to suit the pseudo-classical tastes of the period. It was performed by *lutadores* (wrestlers) in pink tights with diadems and cachesexes of black velvet decked with spangles and little mirrors; and "ladies of the *corps de ballet*", who were men in wigs, cloaks and Roman

[1] For a fuller consideration of the many problems raised by Portuguese and other European ceremonial dances see *The Traditional Dance* by Violet Alford and Rodney Gallop and my articles "The Origins of the Morris Dance" in the *Journ. of English Folk Dance and Song Society* for January 1935, and "The Meaning of Morris" in *The Nineteenth Century and After* for July 1935.

helmets. The former carried clubs and the latter swords. After several unmistakable, if rather sedate, sword-dance figures in which they "heyed" and clashed sword against stave, the *lutadores* formed a human pyramid, like the *torre* in some of the Catalan dances, and the topmost figure, precariously balanced, flourished his diadem in the air.

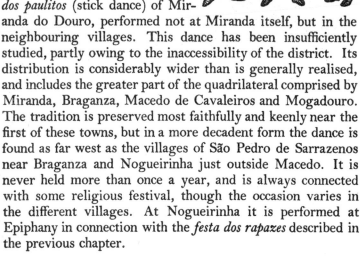

The other performance was a very rustic affair. The young men of Casalinhas, a small village near Torres Vedras, have retained "mauresque" costume but forgotten their steps. They executed *viras*, *verdegaios* and (*horribile visu*) fox-trots, with beribboned wands which must once have been swords, and tall cardboard helmets, pasted over with little pictures in the manner of a child's scrap album.

The most striking sword dance surviving in Portugal is the *dança dos paulitos* (stick dance) of Miranda do Douro, performed not at Miranda itself, but in the neighbouring villages. This dance has been insufficiently studied, partly owing to the inaccessibility of the district. Its distribution is considerably wider than is generally realised, and includes the greater part of the quadrilateral comprised by Miranda, Braganza, Macedo de Cavaleiros and Mogadouro. The tradition is preserved most faithfully and keenly near the first of these towns, but in a more decadent form the dance is found as far west as the villages of São Pedro de Sarrazenos near Braganza and Nogueirinha just outside Macedo. It is never held more than once a year, and is always connected with some religious festival, though the occasion varies in the different villages. At Nogueirinha it is performed at Epiphany in connection with the *festa dos rapazes* described in the previous chapter.

At Cercio, the first village south of Miranda, the dance is associated with a *romaria* to the chapel of Sta Barbara on the last Sunday in August. The *pauliteiros* go round their village some days before, collecting money for the *festa* and, on the day itself, eight or sixteen in number, they perform in front of the church.

When I saw them in 1932 the dancers wore white shirts, black trousers, and waistcoats with coloured ribbons fastened to their backs. From their waistcoat pockets hung on each side a neatly folded white handkerchief. Round their necks were brightly coloured silk 'kerchiefs, the ends drawn through a gold ring on the breast. Their broad, black hats had bands of blue or yellow silk, in which were stuck sprigs of artificial flowers. A generation ago they used also to wear white hand-embroidered skirts and petticoats, and they revived this costume when, as the guests of the English Folk Dance and Song Society, they came to London in January 1934.

Armed each with two short, stout staves, the dancers took up their position in front of the musicians, two drummers and a bagpipe player. The piper droned an introductory phrase, and then broke into one of the many *llaços* or figures of which the dance is composed. With quick walking, running or jumping steps the dancers intertwined to form complex figures, striking each other's sticks with resounding thwacks as they met. Moving from corner to corner, they formed double lines first one way and then at right angles, like the Biscayan sword dancers. Then, as the music quickened, they broke into a spirited "hey". Finally they brought the figure to a close with a movement (common, as it proved, to all the figures) in which, tucking their staves under their arms, they produced castanets to the clicking of which they wheeled into line, while the piper played an attractive little scrap of tune in 6/8 time.

To the lay eye many of the *llaços* were scarcely distinguishable from the first. There were over twenty of them, all named after the words to which, in the curious *mirandês* dialect, the dancers sing them when no instrumentalist is available. *Carrascal, Senhor Mio, Carmelita, A Verde, O Touro, Enramada, A Lebre, A China, O Caballero, O Mirandum* (with words derived from

"Malbrouk s'en va-t-en guerre"), *Vinte Cinco*, *Volticas*, *A
Puentes*, *As Aguias*, *As Bichas*, were but a few of which I took
down the names from their leader, Hannibal do Nascimento
Raposo. Perhaps the most exciting was *O Canario*, in which
the dancers fell apart into groups of four, and two dancers in
each group leaped into the air striking each other's sticks,
while the other two crossed beneath them. In *O Caballero* they
tapped their sticks above and below their lifted legs as in the
Morris Stick dance called "Bean Setting". In
the *Acto de Contrição*, drawn up in two lines,
they dropped for a moment on bended knee before resuming their step, and in the *llaço dos
oficios* (professions) they paused for a moment
to mime the actions of a shoemaker, barber and
carpenter. The whole dance was a remarkable
display of agility and skill.

Local opinion generally contrives to present
such performances as war dances derived from
the famous Pyrrhic dances of the Greeks. Trás-
os-Montes is no exception to the rule. But
spirited and martial as the *dança dos paulitos* is
in character, it is only one link in a long chain
of similar dances running through the Peninsula and all round
Europe, the origin of which is ritual rather than military.

More martial in appearance is the extraordinary performance which, under the promising name of *Mouriscada*, still
survives at Sobrado near Valongo and which retains two of
the elements which have disappeared from the procession of
St John at Braga.

We reached the village of Sobrado at about six o'clock on
the evening of St John's Day in 1932. A village green sloped
down from the high road to a white church backed by a wooded
hillside, vivid in the slanting rays of the sun. A few booths
stood in one corner and there was a general air of rustic revelry.
Making our way down towards the church we found that
the performance had already begun. With drawn swords to
the hilts of which were knotted clean white handkerchiefs the
Mouriscos were dancing in the trellissed courtyard of the

priest's house. They wore light-coloured cotton suits with golden buttons and red belts and sashes. On their heads were cardboard shakos a foot high, hung with little mirrors and gold braid, and surmounted with red plumes. Drawn up in two long lines, with grave unsmiling faces, they were dancing in their places, while their king, distinguished by gold chains and epaulettes, skipped from one end of the line to the other, dancing in turn with different pairs of his men. A comic figure in a blue overall with a mask made of cork bark followed him outside the rank aping all his movements. At one moment the two lines cast off, wound round in a close coil and led off in different directions. The only music was the monotonous beating of a drum which combined with the silence and purposeful air of the dancers to lend their performance a strangely ominous quality such as might presage some outburst of primitive savagery.

When this dance was done the *Mouriscos* marched in single file from the courtyard. At the gate they were met by a mob of figures clad even more incongruously than themselves. The *Bugios*, as these were called, were dressed in gaily coloured cloaks, doublets and trunk hose (hired, in point of fact, from a theatrical costumier at Oporto). They were all masked, a few with animal faces, and on their heads they wore cavalier hats adorned with paper ribbons and culminating in a cascade of paper streamers, like those worn by English mummers. In his hand each carried some household or agricultural implement. They were as rowdy and violent in their demeanour as the *Mouriscos* had been restrained and orderly. With wild gestures and uncouth cries they rushed into the courtyard and began to dance, drawn up like their predecessors in a double line. The music now consisted of a little scrap of melody, interminably repeated by two fiddles and a guitar.

The king of the *Bugios*, who alone was unmasked, took up his position in the centre. He wore a tall plumed shako and an ecclesiastical robe of rich red damask bordered with gold, with a lace flounce round the shoulders. (*The Rei Bugio*, we were told, has the traditional privilege of choosing his dress from among the vestments in the church.) While his follower hopped clumsily in their places, the king summoned each pair of them in turn, and with the swaying gestures of a necromancer appeared to be giving them secret and nefarious orders. Then with an upward sweep of his arms he dismissed the cowering figures. A turn and a flying leap carried them to the end of the line. If the earlier dance had all the character of a preparation for battle, this weird and sinister performance called to mind nothing so much as a Witches' Sabbath, and this on that very day of the year when witchcraft is all-powerful.

Now, the second part of the *Mouriscada* began. Two small wooden platforms had been erected on the village green, some eighty yards apart. In these "castles" the two kings took their stand, each accompanied by a few selected retainers, and began to send mounted "ambassadors" to one another with defiant messages. As each "ambassador" started on his errand with the message in the shape of a scrap of paper or a sycamore leaf impaled on his sword, the warriors fired a salvo with muzzle-loading muskets, and leapt high into the air.

Now it was eight o'clock, and the sun was low on the horizon. There came a moment when the *Bugios* had exhausted their powder and could defend themselves no longer. The *Mouriscos* formed up and marched to the attack of their castle. Twice they assaulted it, and twice they were beaten back. The third time, headed by their king, they swarmed successfully up the ladder. There was a brief *mêlée* on the small platform, and the *Rei Bugio*, masked now, was seen sheltering behind his men. A moment later the Moorish king found him and smote him to the ground. There fell a sudden hush. The fallen monarch sat up, and his followers came in turn to bid him a tearful farewell with tender embraces. Next, with an imperious gesture, his captor bade him descend the steps, the *Mouriscos* formed a close phalanx, and to the strains of a funeral march he was led away. With tears and lamentations the *Bugios* followed behind.

For a moment we were carried far away in space and time. With our own eyes we seemed to be looking at the age-old procession of the dead Osiris. If this was so, then should we not also witness his liberation from the bonds of death? The answer was not long in coming. There was a sudden diversion in a far corner of the green. From beneath their "castle" a party of *Bugios* had dragged out a monstrous "dragon", roughly fashioned of wood and canvas. Holding the *Serpe* like a battering-ram they charged the *Mouriscos* with hair-raising cries. The latter scattered in terror, leaving their prisoner to be rescued by his men and led away in triumph. The dead Osiris had risen again, and now, the mumming over, there remained only the *dança do Santo* to give a Christian finish to this pagan ceremony. Once again the two teams performed their strange dances, this time in front of the church, and in honour of St John, thus bringing to a close one of the most remarkable ritual survivals in modern Europe.

So complex a performance as the *Mouriscada* presents a number of problems to the folk-lorist. In its present form it is clearly an agglomeration of different features, not all of which were always in association. That it is primarily an agricultural fertility rite seems certain. The occasion of its performance affords ample proof of this, and confirmation is furnished by a curious little piece of by-play which takes place earlier in the afternoon. After the *Mouriscos* and *Bugios* have effected their ceremonial entry into the village,[1] but before they begin their dances, one of the *Bugios* rides into the village seated back to front on a pony, and sowing flax.[2] The ground where he passes is next harrowed and then ploughed with two donkeys whose yoke is put on upside down, so that it hangs beneath their chins. Before they reach the village green the plough must fall to pieces. All this, therefore, is done in inverse order and appears to be a charm of "imitative" magic operating by inversion. The *Mouriscos* and the *Serpe* must be

[1] They are met at the gate of the churchyard by the sacristan holding a vessel of holy water which he offers to each king together with a branch of olive with which the latter then asperges his kneeling men.

[2] Though it is flax he calls it maize, although it is hard to say whether any special significance lies behind this confusion of ideas.

inherited from the Braga procession or from some earlier ceremonial which gave birth to both. The former are without doubt a team of sword dancers whose performance is still associated with that mimic death and resurrection, which may well have been the original reason for their existence. The implements which the *Bugios* carry seem to relate them to mediaeval guild dancers,[1] but their uncouth dress and demeanour connect them with other ritual dancers, such as the *Noirs* of the Basque *Mascarades*, while their animal masks and unmasked king recall the *festa dos rapazes* of Braganza. Their name throws no light on the matter: its primary meaning is "monkey", and a secondary one is "clown" or "buffoon".[2]

There remains to be explained the mimic battle between the two armies. It was stated earlier in this chapter that among the dances which have fallen into disuse at Arcozelo da Serra was one simulating a combat between Portuguese and Spaniards. An old British resident at Oporto has told me that in his young days it was nothing unusual in the country round that city, to see bands of mummers playing a battle between Moors and Christians. Similar mock battles are also known in other countries. The Dalmatian *Moreška* is a case in point, and such battles are often the principal feature in the performances of "mauresque" dancers. Local opinion invariably explains them as having been inaugurated to celebrate some victory in wars against the Moors. But it seems far more probable that they are related to the ritual battles between Summer and Winter, an example of which survived till little more than a century

---

[1] In the Corpus Christi procession of 1621 at Oporto "the masons, stone-wers and workers with their King and Standard gave a dance, well costumed, the form of *bugios*, with the instruments of music customarily used in that nce" (Sousa Viterbo, *Artes e Artistas em Portugal*, p. 245). It is worthy of te that Sobrado is close to the Valongo quarries.

The procession held in Lisbon in June 1619 in honour of Philip II of Portugal d III of Spain included dancers representing *bugios e papagaios* (monkeys and rrots). The use of the Spanish word *monos* (monkeys) in a descriptive poem itten by Francisco de Arce on this occasion confirms the above translation *bugios*. (Cf. *op. cit.* pp. 250–5.)

[2] It is worth noting that in this performance we appear to have the two emonial dances distinguished in Arbeaus' eighteenth century *Orchésographie* Morisque and the *Bouffons* or Dance of Fools.

ago as near home as the Isle of Man, and which can only be explained as an agricultural rite of sympathetic magic to assure the triumph of vegetation over decay.

In Soule, the most easterly of the three French-Basque provinces, open-air plays called *Pastorales*, completely mediaeval in technique, are performed in spring or early summer. The good characters are invariably called "Christians" while their opponents, whatever their real nationality, are called "Turks" and wear a traditional scarlet costume with a crown of ribbons, mirrors and artificial flowers resembling the headgear of "mauresque" dancers. In the *Pastorale* of "Jeanne d'Arc" for instance, St Joan and her compatriots are the "Christians" while the luckless English are cast for the rôle of "Turks". Whatever the peripetias of the play, there are always countless battles, stylised to a point at which they become dances rather than realistic representations, in which the "Turks" are finally defeated. Careful consideration has forced me to the conclusion that these plays represent the fusion of a mediaeval theatre of "mysteries" with ceremonial combats between Christians and Moslems.

Something similar seems to have happened in Portugal. In the Azores, according to a writer quoted by Theophilo Braga, the name "Mourisca" is retained for open-air plays on various subjects, while "the historical costume which the actors wear continues to be that of the Moors, as they say". Such a fusion would also seem to furnish the most satisfactory explanation for an elaborate play called the *Auto de Floripes*, which is performed annually on August 5th in the village of Souto da Neves between Barcelos and Viana, in honour of Na Sra da Neves. Like the Basque *Pastorales*, the *Auto de Floripes* combines a mimic combat between Moslems and Christians with a mediaeval play, redolent of the Middle Ages in subject, in stage technique and above all in the blend of stylisation and naïve realism with which it is performed.

I saw it in 1932 on the hottest day of the year. It was to begin at three o'clock, but for two hours we sweltered on the village green with nothing to look at but the stage, which consisted of a long narrow platform with no backcloth other

Plate XI

SINGING TO THE *ADUFE*

GOING TO A *ROMARIA*

Plate XII

AT A PRISON WINDOW

AN OLD BEGGAR

than a stone wall and the green hills on the horizon. At either end of the stage stood a sort of sentry box draped with gaily coloured stuffs. That on the right was surmounted by a white flag with a red cross, the sign of Christendom, while over the other, on a red ground, floated the white star and crescent of Islam.

The first actors to reach the theatre were Charlemagne and his twelve peers. Accompanied by a band, they marched twice round the stage before mounting its right-hand side. The Moors arrived not long after and took up their position at the other side of the platform. They were richly clad in cloaks of purple, crimson and gold, with breeches or skirts fringed with yellow below the knee. On their heads they wore red cylindrical crowns, appropriately serrated in the case of their leader, who bore the strange name of Admiral Balão. The Christians were less splendidly arrayed. Charlemagne indeed wore a crown and a brocade cloak, but his knights were chiefly distinguished by white trousers cut short below the knee, and blue yachting caps. When all was ready a gnarled old drummer took up his stand between the two armies, and the play began.

We found the plot hard to follow, though to the villagers who had seen it every year of their lives it may have been clear enough. The actors either shouted their lines or sang them to an archaic chant. Every important action or gesture was marked by a drum beat or a protracted roll, and the greater part of the play consisted of a series of battles and single combats, of challenges, embassies, and captures, all enacted in a technique so conventionalised as to suggest the ballet rather than the realistic stage. So little has human nature changed through the ages that the writer of this mediaeval *auto* had been obliged, like his counterpart in modern Hollywood, to pander to the popular demand for a love interest. Hence the introduction of Floripes, a beautiful Moorish princess, who formed the apex of an eternal triangle of which Oliver and Fierabras were the other two angles.

There was no questioning the feminine allure of Floripes. The young man who took the part had spared no pains to attain the desired flamboyance. The princess wore a brilliant peroxide wig, a pearl tiara of mammoth proportions, a black bodice with a white lace fichu, and a gaily patterned cretonne skirt. In one

hand she carried a fan and in the other an umbrella. Her arrival in an open victoria created quite a sensation. "Why," people cried, "she really does look just like a girl!"

Floripes was betrothed to Fierabras, a Moorish prince, but at an early point in the play her beauty inflamed the passions of Oliver who, with the help of Roland, carried her off by force. The disappointment of the groom was vividly depicted. His bridal couch had been spread in the very centre of the stage, and there he had lain down, too impatient even to doff his crown. But when he learned that no bride would come to share his princely bed he leapt up in a rage, and there followed a series of battles and challenges, at the end of which Oliver was captured. To the strains of the band the Moors advanced in single file and, like dancers in a *cotillon*, wound round him in a close spiral. Then they uncoiled and Oliver was led away with two Moorish swords pointed at his heart.

There now ensued a series of parleys, ambassadors passing between the two sentry boxes, and finally a truce was arranged. Oliver and Floripes were married: that is. to say, they linked arms, faced opposite ways, and solemnly rotated in the centre of the stage. Then, more realistically, they departed in a carriage, while the bride's father, King Ismar, a comic character, flung sugared almonds among the crowd.

A final battle between the two armies brought the play to an end. Charlemagne and Admiral Balão faced each other diagonally, and, to the sound of a long drum roll, executed a series of twisting sideway leaps, flourishing their swords in the air. Then the band played, and two by two the opposing pairs of knights marched forward and crossed weapons. The Christians fired their muzzle-loaders, and the Moors, armed only with swords, had perforce to surrender. This was repeated until they were all taken, and, like some

children's game of "French and English", the battle came
to an end.

Nossa Senhora das Neves	Our Lady of the Snows
É guía de toda a terra.	Is the light of the world.
Já se renderam os Turcos,	The Turks have surrendered,
Já se acabou toda a guerra.	The whole war is over.

The epilogue was formed by a dance. Charlemagne, Roland
and Oliver performed a sort of "Three Meet" with Admiral
Balão, Floripes, and Fierabras, until finally the whole company
broke into a simplified "Sir Roger de Coverley". It will be
difficult to forget the beauty and colour of that final scene, the
rich robes of the Moors and the slim white figures of the
Christians etched in delicate pattern against the green velvet
of the distant hills.

A less elaborate dramatic form is given to a ceremonial
combat between Christians and Moslems in the obsolescent
Trucos, revived from time to time on the second Sunday in
August at Crasto near Ponte de Lima, at a *romaria* to Na Sra
da Cruz da Pedra. I have never witnessed the performance,
but in the local cobbler's shop I saw the well-thumbed and
scarcely legible manuscript from which it is produced, and
which contains delightfully naïve sketches of the costumes.
The Christians wear bandsmen's uniforms, and the Turks are
distinguished by red stocking caps. Both sides carry shields,
the Turks being armed with swords and the Christians with
lances. Each has a king, a general, a captain and a spy, in
addition to the rank-and-file. After an exchange of challenges,
the Christians speaking their lines in Portuguese and the
Turks in Spanish, there follows a battle in which the Turks
are defeated, and finally, at the instance of an Angel, the infidel
horde is driven pell-mell into the chapel and baptised.

Yet another curious *Mouriscada* is traditionally performed
in the spring by the *maltas* who go down from Beira Alta to
work in the Alentejo. The men sit at a table drinking and
gambling. The Rei Mouro comes and tries to entice away their
women with successively a coin, a song and a flower. Finally
he carries off one of them by force, but he is pursued, and his
castle is besieged and the maiden is rescued.

Folk-plays, both religious and secular, were at one time a prominent feature of church festivals. In the villages round Braganza *autos* are still performed in early summer, the peasants borrowing finery from the town, and dividing their open-air stage in Shakespearean fashion, with notice-boards "This is a house", "This is a forest" and so forth. At Cercio I heard of an *auto* called the *Imperatriz Pursina*. At the same village over thirty years ago Leite de Vasconcelos copied down the text of a home-made *entremes* "*Sturiano i Marcolfa*". This begins with a *Prophecia* or Prologue, of which the first verse runs,

Bou bos cuntar um conto	I will tell you a tale
Para entreter o serão,	To pass the evening,
Bamos falar dos outros	We will talk of other folk
Que os outros de nós falarão.	Since others will talk of us.

In this district there is often a comic character, recalling the Spanish *gracioso* who makes fun of the audience and retails the latest scandals.

In a wood near Chaves, Luis Chaves remembers seeing a pastoral *auto* based on the ballad of the *Conde d'Alemanha*. At a village called Vila do Conde, near the same town, a writer in *Terra Portuguesa* saw a Nativity Play divided into two halves, the first called the *Ramo de fora*, played in the afternoon out-of-doors and beginning with the Creation, while the other, called the *Ramo de dentro*, representing the Nativity, was performed in church after the midnight Mass. A few years later an *Auto da Paixão* (Passion Play), by Antonio Vaz, a sixteenth-century author, was performed at Duas Igrejas (Miranda) with such conviction and realism that the peasant Christ took more than a fortnight to recover from his scourging.

Nativity plays (*autos do presepio*) were widely performed in the churches on Christmas Eve or at Epiphany, and still survive near Vinhais, though about forty years ago they were pro-hibited in Braganza. The *reiseiros* of Maia near Oporto still perform their Epiphany play in mediaeval style, their stage being an open ox cart; and at Eastertide 1932 villagers from the Serra da Estrela hired a small hall at Canas de Senhorim for a Passion Play which was quite mediaeval in its stylisation.

In the eighteenth century ecclesiastical prohibition began almost everywhere to freeze the naïve gestures of the Nativity Play into the immobility of the carved *presepio*, a model diorama in which the central scene of the Adoration provides the pretext for a throng of little figures in contemporary costume engaged in all the diverse occupations of the Portuguese countryside.

Just as the rebirth of Nature in spring and summer was reflected in the folk-festivals of the first half of the year, so also is her lingering death mirrored in the observances of November and December.

The garish *romarias* come to an end in September, and October counts but few *festas*. November brings Halloween and All Souls' Day, called in Portugal the *dia dos finados*, a Christian festival of the dead with visits or processions to the cemetery grafted on to a pagan festival of death which corresponds in the autumnal calendar to the May festival in spring.[1]

[1] A point of resemblance between the two festivals is that in some districts visits are paid to the churchyard on May Day as well as on All Souls' Day.

Two features of the *dia dos finados* reflect its pagan origin. One is the *magusto* or open-air feast of wine and chestnuts (cf. the chestnuts of May morning) probably consumed originally on the graves and undoubtedly connected with funeral offerings. The other is wassailing by bands of children. Round Coimbra these sing *loas* (hymns). At Belas and Obidos they ask for *pão por Deus* (bread for God), a forgotten echo of a traditional charitable distribution, provision for which was made in certain fifteenth-century leases. At Cintra there is a tradition that nothing a child asks on this day may be refused. The carollers are rewarded either with a *magusto* or with *bolos de festa*, special cakes of sugar, cinnamon and sweet herbs.

The evenings draw in. The sun sinks in the horizon until at last the winter equinox comes to mark the end of his long decline, with the feast of Christmas instituted by the Church to supersede an old heathen festival of his rebirth. In the agricultural calendar Christmas corresponds to St John's Day as Halloween to the first of May. There is no intrinsic difference between the festivals of Midsummer and Midwinter except that the first is a communal festival held in the open air, while the other is domestic and held indoors. In Portugal Christmas is essentially a family festival at which as many relatives as possible are re-united.

The bonfires and *galheiros* of June are replaced by the *cepo de Natal*, the Yule log (of oak wherever possible) which blazes in the churchyard or, more usually, indoors on the *fogueira da consoada* (Yule fire) while the family eat the *consoada*, or Christmas banquet in the small hours of Christmas Day. The ashes and charred remains of the log and of pinecones burnt in the fire are preserved and burned later in the year when there are thunderstorms. Where their smoke goes, no thunderbolt will fall.

If the analogy with May Day and St John's Night is to be pursued, we shall expect to find that the beneficent ceremonies inherited from the old pagan festivals of autumn and winter will be accompanied by a belief that certain malign influences have special power at Halloween and Christmas. It is not witches, *mouras* or werewolves however, who are abroad on these two nights. Their place is taken by the souls of the dead, *alminhas a penar*.

In Portugal, at the present day, the *alminhas* appear to be not so much feared as welcomed. Crumbs are scattered for them on the hearth, or the table is left spread after the *consoada*, in order that they may share in its plenty. If they are seen, it is only in the form of little flickering lights. But in the Minho, they come only if no prayers are offered for their peace, and in these prayers it is easy to discern the intention of warding them off. In most primitive communities, the dead are feared little less than witches. Numerous as they are in many countries, the customs in which fear and avoidance are tempered by affection[1] and combined with a ritual aiming at ensuring the return of the dead to their ancestral homes on one day in the year are the exception rather than the rule.

In so far as they are feared, therefore, the souls of the departed may be equated with the malign beings in whom popular superstition has incarnated pagan nature cults.

It is possible that this parallel may prove to be even closer. In the Portuguese calendar the cult of the dead is reflected in other customs which I have not yet described. During Lent, groups of men and women go round the villages late at night and chant the Dies Irae or dirges calling upon all good Christians to awake and pray for the souls of the dead. This is called *aumentar* or *encomendar as almas*. Round Miranda do Douro the song begins:

Acordai, pecadores, acordai, não durmais mais;	Awake, sinners, awake and sleep no longer;
Olhai que estão ardendo em chamas as almas dos vossos pais	Remember the souls of your fathers who burn in flames
Que vos deixaram os bems e vos deles não vos lembrais.	Who left their property to you who remember them no more.

Folk-lorists would probably agree in seeing representatives of the spirits of the dead in the young men with blackened faces, wrapped in white sheets, who run about the country

---

[1] Though even these might be interpreted as the propitiation of beings who are feared rather than as a welcome to those loved.

round Oliveira de Azemeis on St John's Night, howling dismally; and in the figures, also wrapped in sheets, who, hiding
their faces and disguising their voices, break in upon the
desfolhada, the communal stripping of the maize, and are
rewarded with apples by the girls.

There are, indeed, authorities who would go further.
Dr Wolfram of Vienna, for instance, writes as follows: "We
all know the custom of masked men or children going round
at special festivals and begging for gifts. It is remarkable that
the songs in which they ask for gifts are very often not in a
begging tone but importunate and threatening. If they are
given nothing they threaten to throw the chimney down or to
take the roof off, all things that the storm usually does. It is
obvious that the people going round from house to house used
to identify themselves with the storm, the wild-hunt. But the
wild-hunt is no other than the army of the dead."[1]

Portugal appears to offer evidence in support of this theory,
in the association of wassailing with Halloween, and in a verse
from the *janeiras* sung at Covilhã, reproducing the threat to
the house:

Estas casas são bem altas	These houses are very high
Forradas de papelão;	And lined with pasteboard;
Os donos que nelas moram	May the masters who dwell in them
As vejam cair no chão.	See them fall to the ground.

In all ages, a relation has been thought to exist between the
dead and the buried seed from which presently the crops will
spring. The fact that both disappear beneath the earth would
alone suffice to account for this notion, which is reflected in
the myths of Ancient Greece and Rome and in the Greek festival
of the Anthesteria. At this festival, called the "more ancient
Dionysia" and celebrated at the turn of our modern February
and March, the ghosts of the dead rose up, and every householder anointed his door with pitch so that any intruding spirit
should be caught as Brer Fox was caught by the Tar Baby in
Uncle Remus. On the third day of the festival these ghosts
returned to the realm of the dead, taking with them a *panspermia*
or pot-of-all-seeds, which the people entrusted to them that

[1] *Journ. of English Folk Dance and Song Society*, **I**, no. 1, p. 39.

they might tend it and bring it back in autumn as a *pankarpia* or pot-of-all-fruits.

In Portugal the dead are associated with seed, flour or bread in several of the customs mentioned in this chapter; in the crumbs they eat at Christmas, in their intrusion on the maize stripping and in the fact that in some districts the *aumentar as almas* is practised on the threshing floor. In the *terras de Barroso* it is customary, if one drops a crust on the floor, to exclaim: *Para as alminhas* (for the souls).

If, therefore, the dead may be regarded as the guardians of the buried seed and their cult be interpreted as an agricultural fertility cult, the distinction between them and the whole brood of witches and nature divinities appears to dissolve into thin air. No longer does it seem to matter whether the wassailers are to be conceived as harbingers of luck or as the wild-hunt of the dead, since in either event the prosperity of the crops is their aim. If not identical in outward form witches and ghosts are equivalent in so much as both are misty personifications formed in frightened peasant minds by the old rites and their protagonists, divinities once, who have paled to-day into these shadowy creatures of evil.

Now we have followed the year to its uttermost end. There only remains for us finally to dispatch it and so to assure the birth of the New Year in all its young and fruitful vigour. We can choose no better way of accomplishing this than that of the people of Carregal do Sal, who mount their roofs at midnight and thence, with megaphones and appropriate verses, blow away the year that is dead.

Part Three

FOLK-MUSIC AND LITERATURE

"Em quatro versos e oito compasos
Cabe tanta coisa, cabe,
Como só a gente portuguesa a sabe."

(In four short lines and eight brief bars
So much may be expressed, so much...
All that the folk of Portugal can tell,
All that alone to them is known.)

ARMANDO LEÇA:
Da Música Portuguêsa

CHAPTER VIII

THE MUSIC OF FOLK-SONG

In many European countries the existence during the Middle Ages of a native folk-song is no more than a presumption based, at most, upon a few literary references which afford little or no indication of its character. To this rule Portugal forms a happy exception. If we possess no authenticated examples of the songs of the people of Portugal and Galicia (which, from a linguistic point of view were at that period indistinguishable) during the twelfth, thirteenth and fourteenth centuries, we have a wealth of lyric, troubadour verse from which we can deduce the existence of a contemporary native folk-tradition.

The very existence of this poetry was ignored for several centuries. True, as late as 1585 Duarte Nunes de Leão wrote that songs by King Dinis were still extant. In 1621 Antonio de Vasconcellos stated that time had carried them all away; and they were lost to Portugal until the middle of the nineteenth century, when first one and then another of the three great *Cancioneiros* was brought to light. First, in both chronological order and importance came the *Cancioneiro da Ajuda*, then that of the Vatican, and finally the *Cancioneiro Colocci-Brancuti*, so called because it (or one very like it) is known to have been owned, indexed and annotated by Angelo Colocci (d. 1549) and was discovered in the collection of Count Paolo

Brancuti. Of these, only the *Cancioneiro da Ajuda* (which was first printed in 1823 *no paço de Sua Magestade Brittanica, Paris*, by Lord Stuart, British Minister at Lisbon) is contemporary with the poems which it contains, although all three may well have been derived from the collection known to have been formed by Dom Pedro, Conde de Barcelos, son of King Dinis. Together they comprise upwards of two thousand poems, and are known collectively as the *Cancioneiro Geral de Poesia Galego-Portuguesa*.

Hardly any of the poems contained in these codices are anonymous, even the earliest dated example, a poem of 1189, being by a known author, Pay Soarez de Taveiroos. They confirm the many historical references testifying to the existence in Portugal of a cultivated troubadour tradition, the gradual growth of which can to some degree be traced. This tradition, like most important literary movements, was the result of a fusion of both native and foreign elements, and in spite of the adoption of the foreign troubadour system and the deliberate imitation of Provençal verse, it is the native elements which, for both beauty of form and feeling, rank this with the finest lyrical verse of any country and age.

Although there is an intermediate type which owes something to both influences, it is usually possible to distinguish the poems written under the impulse of a native tradition from the more superficial and insipid Provençal imitations.

We are not concerned here with the satirical *cantigas d'escarnho* or *de maldizer* (songs of ridicule and invective) nor with the erudite love poems which were *cantigas d'amigo* and *cantigas d'amor* only in name. These latter names were borrowed from the folk who applied them to their own authentic folk love-songs. *Cantigas d'amor* were the songs addressed by the lover to his lady, and *cantigas d'amigo* were those placed on the lips of a love-sick maiden, proclaiming her devotion to her swain, or more often, with the characteristic Portuguese *soidade*[1] (which is mentioned by name even at this early date), bewailing his absence.

---

[1] *Soidade* or, in its modern form, *saudade*, is an indefinable yearning wistfulness, analysed more closely on p. 262.

These songs, differing from the Provençal imitations both in form and substance so greatly that it is hard to believe that they were composed by the same poets, are known by various names. The most convenient is that of *cossantes*, applied to them in the fourteenth century by Diego Furtado de Mendoza and in the twentieth by Aubrey Bell. No better nor more concise definition of them can be offered than that given by the latter in his introduction to the *Oxford Book of Portuguese Verse*:

They consist of two, four, or more distichs with a refrain, of which the second and fourth, while often altering the sound from *i* to *a* (*pino* o *ramo, amigo* to *amado*), repeat the sense of the first and third, the first line of the third taking up the last of the first, and so on to the end, where the position of the song is found to be very much the same as it was after the first two verses, as far as the sense is concerned. They were dance songs (*bailadas*), danced by the peasants in the villages *de terreiro* or before pilgrimage shrines; *alvoradas* (songs of dawn); pilgrimage songs (*cantigas de romaria*); shepherds' hill-songs (*serranil-as*); boat-songs (*barcarolas*); songs of *ria* and sea (*marinas*). In subject matter, therefore, if not in form, they corresponded closely to the folk-songs of modern Portugal.

Better than any definition, however, is an example, and no better example of the folk-manner of these poems can be given than the first of the seven poems which bear the signature of Martin Codax,[1] *jogral* of Vigo:

Ondas do mar de Vigo	Waves of the sea of Vigo
e vistes meu amigo?	Have you beheld my lover?
Ai Deus, se verrá cedo!	O God that I may see him soon!
Ondas do mar levado	Waves of the sea, uplifted,
e vistes meu amado?	Have you beheld my beloved?
Ai Deus, se verrá cedo!	O God that I may see him soon!
e vistes meu amigo	Have you beheld my lover
por que sospiro?	For whom I sigh?
Ai Deus, se verrá cedo!	O God that I may see him soon!
e vistes meu amado	Have you beheld my beloved
por quen ei gran cuidado?	For whom I am sore troubled?
Ai Deus, se verrá cedo!	O God that I may see him soon!

---

[1] Though "Codax" may well be a name, it has also been suggested that it may equally be a mistranscription of the word "Codex", i.e. the manuscript Martin.

These "parallelistic" strophes with their lines entwined in *leixapren*, as this alternating repetition is called, and the monotony of their alternating endings, conjure up, as few folk-songs do, the formal measures of the round dance. Even without their *son* (music), their rhythm "is so obtrusive that they seem to dance out of the printed page". One seems to see the swaying circles of linked dancers, singing as they go and moving now to the right and now to the left as strophe is followed by antistrophe.

There is, moreover, no lack of literary references to show that "dance or action always accompanied the *cossante*". The very name, indeed, is probably derived from *cosso*: an enclosed place, used for dancing. "The expeditions of Normans and Scots to the *rías* of Galicia", writes Pedro Vindel, "were frequent in the twelfth and thirteenth centuries, and in this part of Spain the arrival of the ships was celebrated with feasts and dancing. According to certain writers, the festivities began with *Cantigas a Santa Maria*, continued with the songs of the pilgrims to Santiago and invariably concluded with the gay songs called *de amor* or *de amigo*."[1] Even when they had been appropriated by the *trobadores*, the *cossantes*, in common with the other poems of the *Cancioneiros*, were still accompanied by dancing. Several of the miniatures which adorn the *Cancioneiro da Ajuda* show the singer accompanied by a girl dancing with castanets or tambourine as well as by the usual instrumentalist with fiddle, guitar, harp or psaltery. When Alfonso VII of Castile went in pilgrimage to Compostela he was received by women dancing *pelas, folias e chacotas*, dances which, if they are unknown to-day, can all be found, accompanied by song, in Gil Vicente's sixteenth-century plays.

Of the three traditional, native features of the *cossantes*, alternating endings, *leixapren* and refrain, all of which betoken their choreographic intention, only the last is generally retained in the modern Portuguese folk-songs, in which, nevertheless, as in the *cossantes*, the dance remains the determining influence. The parallelistic form is still preserved, however, in the dance-songs in Gil Vicente's *Autos*, which are as likely to be pure or

[1] Pedro Vindel, *Martin Codax. Las Siete Canciones de Amor*. Madrid, 1915.

slightly edited folk-lore as the most rustic of the *cantigas d'amigo*.

Towards the end of the last century Leite de Vasconcelos heard songs in the parallelistic form sung at their work by peasants in Trás-os-Montes. And even to-day the last echoes of the *cossante* have not died away, for in 1931 I copied a version of the *Malhão* (a dance-song) from Estremadura, which, though lacking a refrain and more than two couplets, has the alternating endings of the *cantiga d'amigo*:

Malhão, malhão, ó malhão do norte	Winnower, winnower, o winnower of the North
Quando o mar 'sta bravo faz a onda forte.	When the sea is rough the waves are strong.
Malhão, malhão, ó malhão da areia	Winnower, winnower, o winnower of the sand
Quando o mar 'sta bravo faz a onda cheia.	When the sea is rough the waves are full.

In the Portugal of to-day, although the dance still dominates folk-song, the more formal *cossante* has been replaced by *quadras*, single quatrains complete in themselves, which, although they have often inherited the refrain and are occasionally linked together in *leixapren*, have a closer affinity both in metre and in rhyme or assonance with the narrative ballads called *rimances*, *xácaras* or *aravías* described in the following chapter.

The archaic stiffness of the *cossantes*, their freshness and simplicity, free, as only the purest folk-art is free, from intellectual preoccupations or artificial conceits, unmistakably betoken so close and faithful an imitation of folk-art that one is tempted to ask whether they are indeed only imitations. Folk-lore, as we know it to-day, is a creation of the last hundred and fifty years. Seven centuries ago the appropriation of a folk-song with, at the most, a little editing, would not have been regarded as plagiarism. Before the birth of copyright, a signature lacked the full significance which it has, or is supposed to have, to-day. The refrains of the *cossantes*,

in particular, have every appearance of authenticity, ranging
from *Ai madre, moiro-me d'amor* or *E se o verei, velida* to the
simpler *E oj'ei soidade* or *Louçana* and finally to the complete
nonsense syllables (derived perhaps from other languages) of
Alva e vai liero and *Lelie doura, leli leli por Deus leli.*[1] Two of the
poems, João Zorro's *Bailemos agora, por Deus ai velidas* and
Airas Nunez' *Bailemos nos ja, todas, todas ai amigas,* without
being identical, are so similar as to suggest, not that one is an
imitation of the other, but that both are variants of some widely
distributed folk-song. Again, the frequency with which the
same pairs of alternative endings appear suggests a tradition
already old, a long accepted convention and a paucity of
invention which is in the highest degree characteristic of folk-
art, but which would surely not have been tolerated in original
court poetry.

Since it is so abundantly clear that, despite their attributions,
these poems must have been virtually indistinguishable from
folk-songs, it is almost a waste of time to try to prove that they
actually *were* folk-songs. It is sufficient that through them we
know all that we want to know about those folk-songs, except,
unhappily, what is perhaps the most important point of all, the
music to which they were sung.

So complete, unfortunately, is the lack of materials in this
respect, that any theory regarding the nature of this music can
only be founded on hypothesis. Not only the popular *cossantes,*
but also the more artificial Provencal imitations were sung.
That much is certain. Apart from the miniatures in the
Cancioneiro da Ajuda, the first verse of every poem in this
collection is spaced out in a way that could only have been
intended to leave room for a musical notation which, alas, was
never inserted. The only music which we possess is the manuscript discovered in 1914 by Pedro Vindel, containing seven
poems by Martin Codax of Vigo, six of which are accompanied
by music on a five-line stave. These have been transcribed by
Sr. Santiago Tafall, in the style of plainsong.

The contemporary music of neighbouring countries furnishes

---

[1] *leli, leli* may be either a reminiscence of the Muslim creed, or, literally,
"my night" in Arabic.

no satisfactory analogies. That of the Provençal troubadours and of the Castilian *Cantigas a Santa Maria* is not necessarily similar to the Portuguese. Pierre Aubry[1] has put forward the theory that the former is akin to plainsong. It is not probable, however, that the influence in Portugal of Provençal music was any greater than that of Provençal verse. Nor need an independent connection between Portuguese music and plainsong be presumed. True, it has been suggested that the parallelistic form may have been "born in the Church". "The *i*-sound of the first distich...followed by an *a*-sound in the second... may be traced to a religious source, two answering choirs of singers, treble and bass."[2] But Pedro Batalha Reis, who has devoted an interesting monograph to the subject,[3] has shown that a lively secular music flourished at this period, of which little has come down to us for the reason that all learning, or rather all annal-writing, was in the hands of its arch-enemy the Church. The *jograis* (minstrels), who, although sometimes forbidden to compose verse, were its usual vocal executants, and, as trained musicians, may well have been responsible for providing their own music, went, in Sr. Reis' opinion, to popular sources for their inspiration and adapted folk-tunes to the words which their masters had so often borrowed or edited in much the same manner. This appears all the more probable when it is realised that *jograis* must have certainly existed in Portugal before the introduction of Provençal influence. As early as in the reign of Alfonso VII, a Galician *jogral* named Palha is recorded as having been at the Castilian court.

Sr. Reis does not discuss the origins of this secular music. It is quite possible, however, that it was influenced to some degree by the songs of the pilgrims to Compostela, which the Galician folk had so many opportunities of hearing. "The hymns and devotional songs of the pilgrims", wrote Luis José Velázquez in the middle of the eighteenth century, "preserved the taste for poetry in Galicia during the Dark Ages"; and

[1] Pierre Aubry, *Trouvères et Troubadours*. Paris, 1909.

[2] Aubrey F. G. Bell, *op. cit.*

[3] Pedro Batalha Reis, *Da Origem da Música Trovadoresca em Portugal.* Lisbon, 1931.

Carolina Michaelis de Vasconcelos has assessed their influence in the following words:

> The grave, measured, chaste and slightly languid character of the dance songs and, above all, their lack of the licentiousness which might have been expected and which certainly was in existence, may perhaps be explained by the influence of the Church of Compostela which tolerated such manifestations of a religious character, surviving from ancient times, as had never been wiped out from the memory of the people, and transformed them hieratically and liturgically.

Since the *Cantigas a Santa Maria* of Alfonso the Sage were written in Galician, at that period the accepted language for lyric poetry throughout the Peninsula, it has been suggested that their music, of which many examples are extant, may have resembled the Portuguese. For many years all attempts to transcribe this music were founded on the assumption that, like that of the Provençal troubadours, it was akin to plainsong. In 1922, however, Julián de Ribera y Tarragó published his revolutionary work *La Música de las Cantigas*, the conclusions of which may be briefly summarised. Ribera took as his point of departure the fact that in Andalusia the Arabs evolved a form of lyric poetry, which they had not brought with them, called the *zejel* (defined as "a dance song sung in a loud voice before a numerous public"), which not only spread to all Arabic-speaking countries but exerted considerable influence on the lyric forms of all Europe, more especially those of the Provençal troubadours and the German *Minnesinger*. He discovered that some few poems in the fifteenth- and sixteenth-century *Cancionero del Palacio*, published in 1890 by Barbieri, were Spanish versions of Arabic originals, while the verse form of many of them was not unlike that of the *zejel* of Muslim Spain. Next he set out to demonstrate the Moorish character of their melodies (the transcription of which had never offered any difficulties). Acting on this hypothesis he then transcribed the music of the *Cantigas*, and established to his own satisfaction, if not to everyone else's, that it "constitutes a collection of the vocal and instrumental pieces which formed the repertory of the Hispano-moresque professional musicians, at the court of Alfonso the Sage", who was, of course, an enthusiastic exponent of Arabic culture, and wrote the poems in question in newly captured Seville.

It is indeed a fact that Moorish minstrels were so popular throughout Castile that in 1322 the Council of Valladolid found it necessary to condemn the practice of bringing them into the very churches to play and sing at the divine services. But Sr. Reis shows that at no time did this popularity extend to Portugal, and that, whatever influence Arabic music may have had in Castile, it is unlikely to have exerted any direct influence on the music of the Portuguese *Cancioneiros*.

Nevertheless, modern Portuguese folk-songs collected by Correia Lopes in the Upper Douro show pronounced affinities with the music of the *Cancionero del Palacio* and of the *Cantigas*, these two collections being in his opinion virtually identical.[1] He quotes two melodic fragments from the old Spanish music transcribed by Ribera, which he regards as the musical germs, the first (*a*) of his own Douro collection, and the second (*b*)

Ex. 1

of Fernandes Tomás' dance-songs from Beira Baixa.[2] Even more striking is the resemblance which he establishes between the melody of a *rimance* collected at Vila Real (*a*) and No. 242 of Ribera's edition of the *Cantigas* (*b*). These analogies need not, however, be taken to imply that the folk-music of Portugal is of Arabic origin. If, as we have seen, the Arabs of Andalusia borrowed the metres of native poems written in a different language, does it not appear even more probable that they also

[1] Edmundo Armenio Correia Lopes, *Cancioneirinho de Fozcoa*. Coimbra, 1926.

[2] Pedro Fernandes Tomás, *Canções Populares da Beira*. Coimbra, 1923.

borrowed melodies composed in the international language of music? Among the poets who cultivated the *zejel* were some who lived in Portugal: Abengayats of Beja and Abenhabib of

Ex. 2

the Algarve. And at Silves, in the latter province, a Moorish chronicler tells us that "almost every peasant could improvise". It has been supposed that, even if the *cossantes* reflect the "oriental immobility" of the Arabs, the latter could have done no more than restore to the Portuguese a form which was originally theirs. Any analogy, therefore, which may be found between the folk-songs of Portugal and the "Arabic" music of the *Cantigas* may equally reasonably be explained by the theory that the foundation of both is to be found in the native music of the Peninsula, perhaps in that "ululation" to which, according to Silius Africanus, Hannibal's Galicians sang and danced.

The fact is that, although the Portuguese must have sung from all time, folk-music in their country, as in most others, is a blend of many currents, native and foreign, popular and cultivated, secular and religious. In the last resort, all art-music is eventually derived from primitive folk-song. But the invention of melody is assuredly one of the most difficult of all the forms of artistic creation. I am persuaded, though many will disagree with me, that the peasant, left to himself, will invent none but the most rudimentary musical phrases. It is only when these melodic germs are restored to him after having been developed, and expanded in a more cultivated environment where there exist professional or semi-professional musicians, that the gradual process of distortion and modification will begin which is the principal contribution of the folk to their own art, and which gives the folk-song both its anonymity and its distinctive national or local character. To my mind, for instance, the lovely songs of the Hebrides are most probably the outcome of a fusion between the simple occupational songs which the folk created unaided, and the more ambitious music which the harpists evolved from these songs in the mediaeval courts of Scotland and Ireland, the memory of which is retained only in these remote and conservative islands. A similar process must have taken place continuously in Portugal. Just as the blind guitarists of to-day have popularised, in every corner of rural Portugal, the urban *fados*, and, within the limit of my own stay in Portugal, the catchy theme-songs of the "Severa" sound-film, so the *jograis* must have restored to the folk in a more developed form the music which they originally borrowed from it.

If, at the season of the winter sowing, on one of the clear cloudless days of early December, you go out into the Portuguese countryside, even on the very outskirts of Lisbon, you will hear a music which cannot differ greatly from the primitive singing of the Celts of Lusitania. As they guide the oxen with their long goads, the ploughmen intone an endless chant to encourage  the hard-worked beasts. It makes them happy, they say, and unless the oxen are happy they cannot work well. Their song is a succession of simple cries cut abruptly short or with a portando "die-away" like that with which Red Indian songs conclude, and of sustained notes, held for a moment, and then dropping in modal tonality with curious turns, grace notes and flourishes to a tonic a fourth below. Never really different yet never quite the same they are too indefinite to lend themselves to exact notation. They are indeed not so much songs as song in the rough, the raw material, so to speak, of which songs are made.

A curious transition from these rudimentary snatches to the simplest folk-songs is to be found in the *pregões* (cries) with which the street vendors of Lisbon hawk their fish, fruit, vegetables and even lottery tickets. If many of these are no more than raucous cries, or a recitative on two notes, others, within their limited register, are little gems of song, such as the cry of a seller of Setúbal oranges who passed my house every day:

Ex. 3

One day, close by Citânia de Briteiros, the pre-Roman settlement near Braga, I heard a song which itself might well have been pre-Roman. A new road was being built, and some labourers were moving a huge stone which must have weighed several tons. As they heaved and pulled they were encouraged by a self-appointed "shanty-man" who chanted a sort of invocation to the stone:

Ex. 4

So elementary are the tunes of some Portuguese songs that they may well have been evolved from primitive occupational chants without the intervention of any outside influence. Such a one is the song of the pilgrims to Nossa Senhora da Póvoa in Beira Baixa, of which the innumerable variants are all variations of one embryonic musical formula:

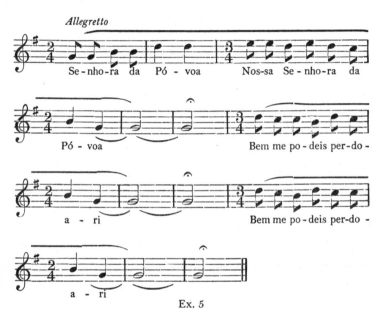

Ex. 5

When one considers the historical and geographical environment in which Portuguese folk-music has evolved, it is easy to understand that instrumental technique is one of the most constant of the influences which have gone to its fashioning. Chronologically speaking, voices came before instruments, and the truest folk-music is, of course, purely vocal. Even to-day most Portuguese songs are sung in unison and unaccompanied; but this is not always the case. Many of the dance tunes are sung to the accompaniment of the guitar, *cavaquinho* (a small mandoline), fiddle, bagpipes, *adufe* (hand-drum), *ferrinhos* (triangle) and drums. The gay dances of the

Douro Valley are often accompanied by a *tuna*, a small band of plucked strings with percussion. The accordion is widely used throughout the country. The favourite combination of Trás-os-Montes is the bagpipe and side-drum, while the *Zés P'reiras*, with their big drums, are a popular feature at many a northern *festa*.

The tonality of the guitar (which was "well-tempered" two centuries before Bach's clavicord), appears, with its alternating chords of the tonic and dominant seventh, to have determined the melodic line of many of the songs. It accounts, for instance, for the emphasis on the leading note, and indeed on the entire melodic structure of this little song from the Alto Alentejo:

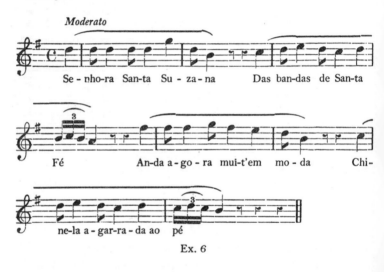

Ex. 6

There is one respect in which the instrumental musician is radically distinct from the singer. Singing is a natural gift possessed by almost all in a greater or less degree, while the player of an instrument is often a professional and always a specialist. He usually has a definite repertoire on which he prides himself and which he seeks to enlarge by adding to it from extraneous sources. Apart, therefore, from any influence he may exert on the tunes which the folk already sing, he is one of the principal channels by which foreign tunes are made known to them and are eventually assimilated to the native folk-song. Portugal has been the centre of a European culture for upwards of a thousand years, and since the time of the *jograis* a class of instrumental musicians has flourished in her towns and larger villages. Little by little their work has percolated among the people, and it is in consequence of these more sophisticated influences that Portuguese folk-music so often has a quality of charming obviousness which may be compared with the countless touches of rustic rococo noticeable in the architecture of the countryside.

If it is admitted that extraneous influences have helped to mould Portuguese song, however, it must be borne in mind that a race will absorb only those things which are congenial to it. Unless, as Ribera supposes, the lively 3/8 time of the *fandango* and other similar dances is of Arab origin, the musical influence of the Moors seems to have been surprisingly small. More than in tonality or rhythm it may perhaps be detected in the dry, toneless, high-pitched voice affected by the peasants of both sexes, although I am by no means sure that this is more than the peculiar quality of voice, intended to carry in the open air, which is indigenous throughout Southern Europe. Equally marked is the relative absence of Spanish influence.

As a general rule, Portuguese songs may be said to lack the remote quality of the Celtic, the poignant reticence of the English, the despair of the Slav and the fire and passion of the Spanish music. They are fresh and charming, unpretentious and infinitely less remote than most folk-music from the music of courts and cities. Imbued sometimes with the cheerful impudence of Tyl Eulenspiegel's *Strassenlied*, they are seldom exuberant, and their most usual mood is one of a gentle melancholy, springing rather from the deliberate cultivation of *saudade* than from any profound grief. They seem to express no stark emotion, but rather one which has been diluted by the soft curtain of rain which sweeps across from the grey Atlantic Ocean. The Northern affinities of this mood are well illustrated in an exquisite little song from Beira Baixa:

Ó pri-mei-ro a-mor que ti-ve   Man-dei o ao ros-ma-

ni-nho   Es-te qu'eu a-go-ra te-nho ai es   tão lin-da   I-

rá p'r'o mes-mo ca-mi-nho

Ex. 7

With few exceptions the Portuguese songs are in the modern major and minor keys. The florid ornament, Byzantine or Arabic in origin, which is so striking a feature of much Spanish music is generally absent. The airs are symmetrical in shape and simple and regular in rhythm. Mixed rhythms and irregular shapes of phrase necessitating changes of time signature are rare. In this the songs reveal their close connection with the dance. Many of them, indeed, fall into two distinct halves: the first a slow narrative section, and the second livelier and more obviously choreographic. While the first section is sung the dancers walk round in a circle with linked hands, and the quicker measures of the dance coincide with the refrain.

A good example of these composite songs is a *Malhão*. Two dance-songs, the *Malhão* and the *Ladrão* (thief), are widely distributed over Portugal in countless different variants. Francesco Lacerda held that they stand in the folk-imagination for the flail of hearts and the thief of love. The *malhador* (*malhão*), who at harvest time goes from farm to farm earning a living as a casual labourer, may often leave more than a garnered harvest behind him, and it is easy to see how he may have come to be the prototype of the man "who loves and rides away":

Andante

Ó ma-lhão   tris-te ma-lhão

Ex. 8

O winnower, sad winnower,
A sad life will I lead thee.
Neither will I wed with thee,
Nor let thee marry.
Mas o ai la-le-lo-le-la.
Nor let thee marry.

The words support this view as does a version of the *Ladrão*, collected at a wedding-feast at Serpa:

Fôs-te tu fôs-te tu la-drão Fôs-te tu fôs-te tu la-drão rou-ba-dor das chaves do meu co-ra-ção rou-ba-dor das chaves do meu co-ra-ção

Ex. 9

It was you, it was you, the robber
Who stole the keys of my heart.

Although they all bear a strong family likeness, the Portuguese dances are innumerable. Some of the names are peculiar to certain districts such as the *Corridinho* to the Algarve, the *Fandango* to the Alentejo and Ribatejo, the *Bailarico* to Estremadura, and the *Rusga* and *Rabela* to the Douro. Other names are generally used throughout the country, although the airs and dances to which they are applied are by no means identical. Thus the *Verdegaio* of the maidens of Viana is quite different from that of the scarlet-waistcoated *campinos* of Vila Franca da Xira, nor is their *Tirana* the same as that of the check-shirted fishermen of Ovar. Of the *Vira* there are as many variants as there are districts.

Most of these dances are founded on a more or less varied *Jota* step and on simple country-dance figures. "In many societies, particularly in the provinces," wrote A. P. D. G. in 1826, "the English country dances are still in use." The *Pretinho* of Carreço is nothing more than "Stripping the Willow" under another name, and the *Tirana* and *Pai do Ladrão* danced in the same village might also be English country-dances. The *Jota* step is executed opposite partners in the *Fandango* and *Vira do Minho*. Both step and figures are most gracefully blended in the *Viras* of Central Portugal and other dances performed in fours or eights. There is much swaying of bodies and snapping of fingers in castanet fashion. The general impression is one of airy gracefulness, in strong contrast with the sculptural arrogance of the Spanish dances.

In the dance-songs with a slow solo section and a livelier refrain, the former was often originally derived from one of the old airs to which the narrative ballads were traditionally sung. These melodies are usually in the minor key and are more archaic in character than the dance refrains. The following is one of the rare examples which I was able to collect:

Andante

E-stan-do a be - la In - fan - ta   No seu jar-dim   do

ma - ri   Com pen-te d'oi-ro na   mão

Seu ca - be-lo pen - te - a - - va
Ex. 10

Since, as will appear, the ballad form is believed to have come from Spain it is not unreasonable to presume that its music had the same origin. Indeed, the ballad airs are very similar in both countries. The Spanish *romance* was derived from the old *cantares de gesta* with their uneven line of from eleven to twenty syllables. Of these J. B. Trend writes in *The Performance of Music in Spain* that "practically every line has four down-beats and four up-beats, a system probably derived from folk-dancing". Thus if, as may well be supposed, the music of narrative songs grew out of that of the *cantares de gesta*, it is ultimately derived, no less than the livelier refrains, from the dance.

Though the music which we have considered up to this point is characteristic of the country as a whole, it is only in the Western districts that it is heard to the exclusion of all else. In the East where Portugal marches with Spain, on the open horizons of Trás-os-Montes, Beira Baixa and Alentejo this music rubs shoulders with another freer in melodic line and rhythm, deeper and more profound in its emotion. The songs sung in a curiously free three-part harmony at Serpa or to the rhythmic accompaniment of the *adufe* round Penamacôr and Idanha, express the severe dignity of the landscape as faithfully as those of Minho and Estremadura conjure up the green woods and fields of the West.

To some slight extent the proximity of Spain may account for this difference. In the music of Idanha and Elvas, in particular, Spanish influence may be discerned. The thrilling pilgrimage song of Na Sra do Almurtão from the former district ends in the "Phrygian" cadence characteristic of Southern Spain:

Ex. 11

So does a charming song to which people were dancing at the
Póvoa *romaria* and which is written in the Spanish *guajiras*
rhythm, a compound of 6/8 and 3/4:

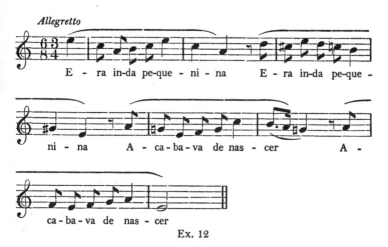

Ex. 12

But Spanish influence does not go deep, and the *tiempo de
guajiras* is more often found in combination with melodies of
a more characteristically Portuguese curve:

Ex. 13

There is nothing Spanish in the rhythm of a St John's song
from Covilhã:

Ex. 14

or in *Meu Lirio Roxo*, the most beautiful of the songs I heard
at Serpa, the flattened seventh of which suggests the influence
of the Mixolydian mode (G to G on the white notes):

me a-te-mo - ri - - zo   Eu   de - la   não

me a-te-mo - ri - - zo

Ex. 15

Indeed in these remote Eastern districts modal tunes are by no means rare, and many of them are of great beauty. Two reaping songs, the one from Caseigas near Covilhã and the other from Meimôa near Penamacôr are respectively in the Dorian (D to D) and the rare Lydian (F to F) modes:

Adagio

Ó que cal-ma vai ca - in - do   Ai!   os cei -

fa - do-res do   cam - po   Meu a-mor por lá an-

das   Ai!   en-cos-ta-do ao li - ri - o bran - co

Ex. 16

Andante

Ai! s'eu te de - ra da - va   da - va 'sta - va sem-pre a
Bei - ji - nhos a - té   mor - rer - ri   E a - bra-ços a -

dar   e da - ri
té a-ca - ba - ri   Ex. 17

For all its final Phrygian cadence the Spanish scale above alluded to is not a true Phrygian mode (E to E). It is to be regarded rather as a peculiar form of the minor scale ending on the dominant, a view which alone accounts for the occasional sharpening of G the seventh note of the scale by attraction to the tonic A. In Eastern Portugal, however, the true Phrygian mode is also found, and two examples of it are appended:

Ex. 18

Ex. 19

These modes need not necessarily betoken an ecclesiastical
influence. In England, for instance, they definitely do not do
so. In Portugal, however, the circumstantial evidence in favour
of such an influence cannot be neglected. It has been mentioned
that the songs of the pilgrims to Compostela may have been
among the sources from which Portuguese folk-tunes derive.
Among these are to be found one or two with the austere simplicity and the slow, firm beat which one would associate with
the chants of marching pilgrims, as for example the *Alviçaras*
sung at Penamacôr on Easter Eve:

Ex. 20

A St John's song from the Serra da Estrela has not only all
the character of plainsong but also retains on the word *ver* the
"hocket" (pause for breath) of mediaeval ecclesiastical chant:

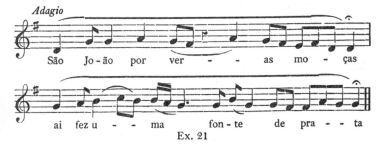

Ex. 21

The folk-songs of Portugal, and in particular the two divisions into which they fall, leave many problems to be solved. Is the music of the West to be regarded as of more modern development and derived from more sophisticated and largely instrumental sources? Is that of the East the last vestige of a more ancient vein of song preserved partly by geographical isolation and partly by the atonal accompaniment of the *adufe*? Or are not both equally indigenous to Portugal? If the latter view is accepted, then the twin branches may perhaps be derived the one from secular and the other from ecclesiastical roots. The use of modes and free rhythms is the argument in favour of such an origin for the music of the East. The fact that that of the West is mostly in the major key need however be no bar to its antiquity. For the modern major scale is no other than the ancient Ionian mode, the *modus lascivus* which Gregory refused to admit into his system of ecclesiastical music, on the ground that it was the one in which the "ribald ballads" of the people were cast. In an article on "Music in Spanish Galicia" published in the first number of *Music and Letters*, Professor J. B. Trend demonstrated the existence in the north-west of the Peninsula, as early as the sixth century, of a profane, pagan music which incurred the displeasure of St Martin of Braga and the use of which in Christian Churches was prohibited by the Council of Lugo in 571. What more likely, then, that the "lascivious" major was the mode not only of the profane music of Portugal and Galicia, but also of the aboriginal Spanish music before it was affected by Eastern influences?

In the midst of so much uncertainty the most that can be hazarded is a guess that Portuguese folk-music has always had something of the character which distinguishes it to-day, since the folk tend to assimilate only such airs as conform to their taste, while to such as do not do so they remain ever impervious.[1]

---

[1] With the exception of Nos. 1 and 2 the musical examples quoted in this chapter are taken from my collection of 130 Portuguese folk-songs which is shortly to be published in its entirety in Portugal. Nos. 7, 9, 13, 16, 17, 18 may be found with piano accompaniments and English singing versions in my *Eight Portuguese Folksongs*, published by the Oxford University Press and Messrs Sassetti of Lisbon.

CHAPTER IX

THE TRADITIONAL BALLAD

It is seldom nowadays that one has a chance of hearing the old traditional ballads called *rimances*. Gone beyond recall are the times when they were sung round the fire during the long winter evenings or in the fields to beguile the arduous days of harvesting under the scorching August sun. True, there are still blind, wandering minstrels of the kind who used to sing them and sell broadsheets that all might learn. But to-day they are purveyors only of doleful *fados* chronicling the latest sensational crime or of popular successes from the revues and sound-films of the capital. Only rarely, in some remote village encircled by mountains, buried among sea dunes or lost in the vast spaces of the plains, does one stumble upon old people who, like the poet Almeida Garrett's nurse Rosa de Lima, can still chant to some ancient, dignified tune the legends of *Dona Alda* or *Gerinaldo*.

For some reason or other these Portuguese ballads are less widely known than those of other countries such as Spain. This may be partly due to the fact that the ballad was not originally a native form in Portugal. It appears to have been imported from Spain in the sixteenth century when Spanish influence was predominant at Lisbon. At that time the two languages, although already distinct, were less completely divorced from one another than they are to-day, and throughout the Peninsula Castilian was the accepted language for epic and dramatic literature, as Portuguese was for lyrical and love literature.

Nevertheless, the Portuguese genius made of the ballad something very different from the Castilian, and in its way no less beautiful. Although they were not chosen from the stirring episodes of Portuguese history, the themes of the *rimances* are almost always distinct from those of the Spanish *Romancero*. It has even been suggested that many of the latter which have something of the lyrical, imaginative quality of Portuguese literature (such as the lovely *Conde Arnaldos*, for instance) may have been written in the Castilian tongue by Portuguese ballad-mongers. Some of the finest Portuguese ballads deal with the sea, and among them is one, *A Nau Cathrineta*, in which may be seen the original both of a well-known Basque folk-song[1] and of our own comic *Little Billee*:

There sails the good ship *Catherine* of which I now will tell.
Listen, gentlemen, if you would hear a very wondrous tale.
For more than a year and a day she had sailed upon the sea.
There was nothing left to eat; there was nothing left to drink.
They set shoe-leather to soak, to sup on the following day;
But the leather was so tough that they could not swallow it.
They cast lots to fortune, which of them should be killed,
And the lot fell upon the Captain-General.
"Climb, climb, my sailor lad to the top of the main mast,
See if you can descry lands of Spain or sands of Portugal."
"I see nor lands of Spain nor sands of Portugal,
But seven naked swords ready to slay you dead."
"Up, up, topman, to the main crow's nest,
See if you can descry lands of Spain or sands of Portugal."
"Give me a reward, my captain, my Captain-General.
I can see the lands of Spain and the sands of Portugal:
And I can see three maidens sitting beneath an orange tree,
One of them sitting sewing, another spinning on a rock,
And the fairest of the three is between them weeping."
"All three are my own daughters. Oh that I might embrace them.
The fairest of them all I will give to you in marriage."
"I will not have your daughter who cost you so much care."
"Then I will give you so much money as never can be counted."
"I will not have your money which cost you much to win."
"I will give you my white steed always ready to gallop."
"Keep your horse which cost you care to train."
"Will you have the good ship *Catherine*, to be her captain?"

[1] Cf. *A Book of the Basques*, p. 114.

"I do not want the *Catherine*, for I know not how to sail her."
"What do you want then, my topman? What reward shall I give you?"
"I only want your soul to carry off."
"I renounce you, O Satan, you that have tempted me.
My soul belongs to God, and my body shall rest in the sea."
An angel took him in his arms and would not let him drown.
The Devil burst, and the sea forthwith was calmed,
And that night the good ship *Catherine* came safely to the shore.

Most numerous among these Portuguese narrative poems are the ballads of chivalry. "All the world loves a lover", and the Portuguese, who make a virtue of sentiment, do so more than most. It would be no exaggeration to trace this attitude back to the troubadour tradition which flourished in the thirteenth and fourteenth centuries. The Breton cycle of legends dealing with Tristan, King Lear and the Round Table were well known and much appreciated. Many Portuguese nobles of the fourteenth century bore the names of their heroes, and Nuno Alvares the Holy Constable, Portugal's *chevalier sans peur et sans reproche*, modelled himself upon Sir Galahad and, in the words of Aubrey Bell, "came as near realising his ideal as may be given to mortal man".[1] The Portuguese versions of these legends have been lost, but the ideals derived from them held sway at Court for many years and inspired a literary form, the "romance of chivalry" of which the prototype *Amadis of Gaul* was spared by the Priest and the Barber in *Don Quixote* when they consigned to the flames its many imitations. Ridiculous and unreal as were these novels, they rendered the country a signal service in propagating loftier and more chivalrous ideals among ever-widening circles of the folk.

It was at about this time that the ballad made its appearance in Portugal. What more natural, then, than that the humble people, hearing from afar the echoes of glittering Court life, their imaginations aflame with its glamour, should have been avid of any tale that would bring it nearer and afford them some escape from their prosaic daily round.

The ballad of *Dona Infanta* is not only one of the most popular, but one of the most characteristic in its expression

---

[1] Aubrey F. G. Bell, *Portuguese Portraits*, p. 62.

of the two principal ideals of the Breton *lais*, the loftiness of love and the value of woman's virtue:

The beautiful Infanta was sitting in her garden,
With a comb of fine gold she combed her hair.   ·
She cast her eyes out to sea, saw a fine armada sailing,
The Captain who sailed therein, right skilfully he steered.

"Tell me, O Captain of your noble armada,
Saw you my husband in the land the Lord's feet trod?"

"So many knights are in that holy land...
Tell me O lady, the signs by which men knew him."

"He rode a white horse with a saddle of silver gilt.
At the point of his lance he carried the cross of Christ."

"By the signs that you have told me, I saw him, in a skirmish,
Die the death of a hero. I avenged his fall."

"Ah woe is me, poor wretched widow that I am
With none of my three daughters wedded."

"What would you give, O lady, to the man who brought him hither?"

"I would give him gold and fine silver, and all the riches that are mine."

"I do not want gold and riches, I have no need of them.
What more would you give, O lady, to the man who brought him
    hither?"

"Of the three mills that I own, all three would I give to you.
A fine flour they grind, the King took it for himself."

"I do not want your mills, I have no need of them:
What more would you give, O lady, to the man who brought him
    hither?"

"The tiles of my roof which are of gold and marble."

"The tiles of your roof, I have no need of them:
What more would you give, O lady, to the man who brought him
    hither?"

"Of the three daughters that I have, I would give to you all three,
One to shoe you, one to clothe you,
And the fairest of all to sleep with you."

"Your daughters, O Infanta, are not the maids for me,
Give me something else, O lady, if you would I brought him hither."

"I have no more to give you, and you no more to ask."

"Nay, O lady, not yet have you offered me yourself."

"The knight who asks this thing, who is himself so vile,
Seized by my own villeins I will make him go
The round of my garden tied to my horse's tail.
Vassals, O vassals, hasten quickly to my side."

"This ring with seven stones which I divided with you,
Where is its other half? My own half have I here."

Since, however patrician their origin, these ballads became
with time the exclusive property of the folk, who naturally
assimilate only what is acceptable to them and reject what is
not, the conception of love which they reveal is a measure of
the remarkable extent to which the ideals of the Court pene-
trated to the lower orders of society. It is summarised most
neatly in the conclusion of the ballad of Count Alves:

Apartar os bem casados era o que	To separate those truly-wedded
Deus não queria.	was against the will of God.

Implicit in all these poems is the blind, desperate, unreasoning
love-at-first-sight of fairy-tale rather than the more serene if
no less radiant love of real life.

In the ballad of Dom Martinho de Avizado an old knight
bewails the fact that he is past the age for bearing arms and
that his seven children are all girls. His eldest daughter asks
him for arms and a horse, and offers to go in his stead to the
wars between France and Aragon. The father objects in turn
that her eyes are too bright, her shoulders too high, her breasts
too firm, her hands too small and her feet too delicate. Men
will know her for a woman. She replies that she will keep her
eyes lowered, the heavy armour will weigh her shoulders
down, and a well-shaped breastplate will conceal her breasts,
iron gauntlets her small hands and spurs her shapely feet. The
scene then changes abruptly, and a young man is heard complaining to his father and mother that "there is a great pain
in my heart, for the eyes of Count Daros are no man's eyes but
a woman's". On his father's advice he tries to trick "Count
Daros" into betraying her sex. The tests differ in the many
variants of the poem. The knight asks "Count Daros" to dine,
hoping that she will give herself away by sitting down on the
ground or by breaking bread against her breast (a curious light

on the manners of the period). He leads her into the garden, hoping that she will smell the flowers. He takes her a'fairing counting on her to fly to the ribbons:

> Oh what pretty ribbons to adorn a lady.

But "Count Daros" is interested only in manly things:

> Oh what a fine dagger is this to fight with men:
> Pretty ribbons are but for ladies' wear.

Finally the knight challenges her to go swimming with him. This she evades by inventing a message that her mother is dead and her father dying and in need of her. But the tale ends happily, for after seven years of warfare she returns to her father offering him as son-in-law her captain whose love alone was keen enough to pierce a disguise defended with such determination.

One of the most beautiful lyrical passages in Portuguese poetry occurs in the ballad, *O Conde Alberto*, in which a king's daughter falls so passionately in love with a count that the king orders the latter to kill his own wife that he may marry the princess. The knight returns home sad at heart but convinced that his honour requires him to obey the royal command. His wife argues with him in vain. He must offer up her head in a golden basin, a reminiscence perhaps of John the Baptist's charger. Before she dies the countess bids farewell to the world:

> It is not death but jealousy that makes me sad,
> And sadder still, Lord Count, your cowardice
> To slay me merely at the King's behest.
> Let me now bid farewell to all that I have loved,
> Let me go from the hall into the garden:
> Farewell carnations, farewell roses, farewell flower of rosemary.
> Farewell serving-maids, farewell my women, with whom I played
>   so happily.
> To-morrow at this very hour I shall be in the cold earth.
> Give me my little son, that I may suckle him.
> Drink, drink, my little son, this milk of bitterness,
> To-morrow at this very hour thy mother will be in her grave.
> Drink, drink, my little son, this milk of agony,
> For thy mother is to die who loved thee so well.

Beautiful, too, is the farewell message spoken by the ghost of his wife to the unfaithful Bernal Frances in the ballad of that name:

> Live thou, dear love, live thou for I have lived:
> The arms which once encircled thee no longer can embrace:
> The lips with which I kissed thee have lost their savour:
> The eyes with which I looked on thee behold thee no more.

The ballad in which love is carried to its highest apotheosis is the short but exquisite *Conde Ninho*, in which a motive occurs which is familiar in our own border ballads:

> There goes the Count, Count Ninho; he goes to water his horse:
> And while the horse was drinking he sang a song so clear.
> "Drink, drink, my horse, for God will free you
> From the labours of this world and the sands of the sea."
>
> "Awake, O beauteous princess, I have heard so sweet a song:
> It is either the angels in heaven or the sirens in the sea."
>
> "'Tis neither the angels in heaven, nor the sirens in the sea,
> 'Tis the Count, Count Ninho, who wants to marry me."
>
> "If he wants to marry you I will have him slain."
>
> "When you put him to death, order me to be beheaded:
> Let me be buried at the door and him at the foot of the altar."
>
> The one died, and the other died, and there they go to bury them:
> From the one sprang a little pine tree and from the other a fair pine:
> The one grew and the other grew, and their branches became entwined.
> When the King went to Mass they would not let him pass.
> Wherefore the accursed King ordered them to be cut down:
> From the one ran pure milk and from the other royal blood.
> There sprang from the one a pigeon and from the other a ring-dove.
> Sat the King at table, they perched upon his shoulder:
> "A curse on so much loving, a curse on so great love;
> Neither in life nor in death could ever I sunder them."

A corollary of this idealisation of love and of chastity is the savage ferocity with which infidelity is punished. The penalty of impurity is clearly stated in the ballad of the *Captive Count*:

> Da lei divina é casar-se, da humana ser degolada.
>
> Marriage, by the law divine, and by the human, death.

Only the premature demise of the culprit, in another ballad, saves him from exculpating his crime under both codes.

It is easy, therefore, to understand the alarm of Dona Ausende, in the ballad of that name, when she finds herself with child through having eaten a magic herb growing at the door of her father's palace. In vain she seeks to disguise the fact by pretending that her clothes are ill-cut. She is condemned to be burnt alive, and from this fate she is saved only by the timely discovery that the same herb which makes virtuous women pregnant has the opposite effect on expectant mothers.

The most vigorous and dramatic of these tales of adultery in high places is the *Conde d'Alemanha*:

The sun rises behind the hills and the clear day is here
While the Count of Germany was still sleeping with the Queen.
The King knew it not, nor all the courtiers,
Alone in knowing it was Princess Juliana, his daughter.
"Juliana, if thou knowest it, tell it not;
For the Count is very rich and will clothe thee in gold."
"I do not want his robes of gold, I have enough of damask;
While my father yet lives you would give me a stepfather!
The pleats of this gown, may I never fold them
If when my father comes from Mass I do not go and tell him all,
The pleats of this gown may I never finish them
If when my father comes from Mass I do not tell him everything."

Upon these words her father knocked on the door:

"What talk is this between a mother and her daughter?"
"Welcome my father; for thy coming God be thanked;
I have a marvel to relate to thee:
While I was at my loom weaving fine cambric
The Count of Germany came..." "And broke your thread:
Do not be angered, daughter, nor yet anger me,
For the Count is merry, perhaps it was a jest."
"A curse upon his jesting and upon his black drollery;
For he took me by the arm and would have bedded me."
"Then calm yourself my daughter, and give me no more cause for anger,
To-morrow at this very hour the Count's throat shall be cut."
"Arise, my mother, and behold a wondrous sight,
'Tis the Count of Germany, a company go with him,
With his head in a basin and his blood in a bucket."

"Thou hast done ill my daughter to the milk that thou hast sucked,
So fine a Count, and thou hast done him to death."
"Calm thyself, mother, and give me no more to grieve,
The death the Count has met, let me not see thee meet it."
"Thou hast done well, my daughter, to the milk that thou hast sucked,
A maid of twelve years old, thou hast saved me from death."

Against these violent and melodramatic themes may be set
others in which a native mother-wit has parodied them.
Maidenly honour, usually so firmly defended or so savagely
avenged, in these is tricked into a fatal moment of weakness
which the ballad-monger, as though with a sly wink, seems
actually to condone. If in the ballad of *Bela Infanta*, the knight
returned from the wars finds his wife constant to his memory,
the soldier in the ballad *O Soldado* finds a different state of
affairs in his home. Granted seven days' leave, on the condition
that he may take seven years if he finds his wife bewailing his
absence, he discovers her in another man's arms and unwilling
so much as to open the door to him. In disgust he returns to
his regiment:

> With seven days of leave, seven hours are more than enough
> To free me of *saudades* and to cure me of my longing.

The spirit of satire is again uppermost in a ballad, the
subject of which was probably taken from one of those "merry
tales" which were so popular in the Middle Ages. A knight
hunting in the woods finds a maiden who has been bewitched
"whilst yet in her mother's lap" so that she must remain in
a tree for seven years and a day. One of the Azores versions
shows well what manner of maid she is:

> With a comb in her hand with which she combed herself,
> The hair of her head flowed over all the tree,
> The eyes in her head brightened all the world,
> The apples of her cheeks were as fine rubies,
> The teeth in her mouth resembled fine crystal,
> And from her lips ran scarlet drops of blood.

To-day the spell comes to an end, so the knight gallantly
escorts her out of the wood, without seeking to take advantage
of her helplessness. (In one version he is not quite so chivalrous, but she deters him by saying that through a spell which

has been laid upon her the man who becomes her lover shall be turned into a mulatto.) When they come to the edge of the wood the maiden laughs. The knight asks why. "I am laughing at a starling which flew chattering through the air", she replies. A little farther on she laughs again, and this time she confesses:

> I am laughing at the knight, and at his cowardice,
> Who had the maiden at his mercy, and never took her.

Too late the knight suggests that they should turn back into the wood to look for a silver spur which, he says, he left at a fountain where they stopped to drink. "If you have lost a spur of silver, my father will give you one of gold," says the maiden, and proclaims herself to be the King of France's daughter.

Some of these poems, notably those from round Coimbra, are in the form of dialogues between a young man and a maiden who counters his declaration of love with artful evasiveness, a *genre* which is also still very much alive in the Basque country.

The burlesque character is best shown in the *Xácara da Moreninha*:

> I went to the door of the dark-eyed fair, of Morena the ill-married:
>
> "Open the door to me, Morena, open to me for your soul's sake."
> "How shall I open the door to you, Father John of my soul,
> With my child at my breast and my husband at my side?"
>
> During this colloquy, the husband awoke:
>
> "What is this, my wife? To whom do you address these words?"
> "I am talking to the over-boy, who came to see whether they were kneading,
> I told him, if they knead milk-bread, to add but little water."
> "Arise, my wife, go and see to your house.
> Send your lads for wood and your slaves to fetch water."
> "Get up yourself, husband, go and shoot in the mountains,
> There's no better rabbit than the one that's shot at dawn."
> The husband went out, and Morena bedecked herself,
> With her scarlet cloak at twelve testoons a yard
> And stockings of red silk stretched tight upon her leg,
> Her stick in her hand which scarcely touched the ground.
> She went straight to the monastery and came to the gateway.
> Father John who saw her coming, leapt rather than ran,
> Took her to his cell and confessed her right well.
> The penance which he gave her, forthwith she performed it.

Plate XIII

OIL JARS AT ESTREMOZ

Plate XIV

On leaving the monastery her husband met her.
"Whence come you, wife, whence come you so well satisfied?"
"I come from hearing Mass, a new Mass well sung.
The Celebrant was Father John, therefore am I well satisfied."
"I'll satisfy you with the point of the sword!"
He gave her a thrust between the breasts that left her stretched out
    dead.
"It irks me not to die, for dying costs nothing;
It grieves me for my little daughter that I leave her unweaned."
"Were you a better mother you had not been so ill-wedded,
Nor would you have to die this tragic death."
They carried her to the convent and laid her in the tomb.
Father John smiled: and it was the husband who wept.

The various themes of the *rimances* having been reviewed,
it is now possible to venture upon some more general conclusions regarding their origin and character. That, as collected
in the nineteenth century, they are informed with a very real
folk-quality is undeniable. Apart from the circumstances in
which they were sung, their form and mannerisms alone are
sufficient proof of this. One of the most characteristic mannerisms of the folk-style is economy of invention. This is
exemplified in these poems by the frequent reiteration of
certain well-worn formulae figuring most usually at the beginning or end. A conventional opening such as

Manhaninha de São João, pela   On the morning of St John, so
    manhã d'alvorada.   early in the morning.

may serve as introduction to half a dozen different themes, and
can be compared to the English "As I went out one May
morning, one May morning so early". Again, the same words
and expressions will always be used to describe the same or
similar incidents, even when they occur in different ballads.
In no country are monotony and redundancy regarded as tedious
by the ballad-singer or his audience.

It is clear, also, that these poems were meant to be heard
rather than read. The personality of the singer is always
implicit in them, illustrating the tale with modulations of voice
and gesture, playing all the rôles in turn. Hence their vividly
dramatic quality and the elliptical and circumstantial style with
its abrupt excursions into dialogue. The story is enacted and

commented rather than told. Here and there a minor incident
is described in the most minute detail. The rest is left to the
imagination or previous knowledge of the listener.

The corrupting influence of oral transmission is everywhere
apparent, despite the assistance which metre and assonance
afford to peasant memories. For instance the opening line

Mi madre era de Ronda y mi padre de Antequera,	My mother was from Ronda and my father from Antequera,

of a well-known Spanish ballad appears in Portugal as

O meu pai era de Hamburgo, minha mãe de Hamburgo era.	My father was from Hamburg, my mother from Hamburg was.

Different legends are often confused, and new ballads arise
from their fusion. The examples given in this chapter have
been chosen for their clarity and completeness (although those
collectors have been avoided who are suspected of having
obtained such "perfect" versions only by the synthesis of
several incomplete variants). It would have been easy to
replace them with others in which all continuity of narrative
and consistency of motive are lost.

Some of the *rimances* indeed survive only in suggestive
fragments such as the following:

"I would give thee my horse and all the money I carry,
Let me go to my house to take leave of my aunt."
"I do not want thy horse, nor aught from thy hand;
All I want is to slay thee and to pluck out thy heart."
Three days he lay there and none knew aught of it,
They learned it only from the birds that went to pick his bones.

But if the poems reveal much of the folk-style, their themes
are usually far removed from the ken of the peasantry, and it
is scarcely conceivable that the folk can have been the inventors
of the stuff from which they are fashioned. The lack of historical
ballads has already been mentioned, and in this connection the
shortness and unreliability of folk-memory is notorious. From
Torres Vedras to Busaco, for instance, all recollection of the
Peninsular War has faded in little more than a century. Nor
does it appear that the Portuguese peasants, despite their gift
for improvisation, were ever addicted to weaving into ballads

the humble tragedies of their own experience. Where such themes appear, their handling is trite and unimaginative:

> In an old castle lived Dom Pedro
> With his wife Joanna Maria.
> He was strong and sturdy, though old,
> And in his spare hours he played the clarinet.
> One fine day they came and told him
> That his Joanna was false to him.
> So then Dom Pedro sets himself to watch
> And he sees a fellow walking about.
> He goes running up to his room,
> Fetches a musket and returns in a moment.
> Blind with fury he aims...and then
> He fires the gun, Pum! and kills the two of them.[1]

When, on the other hand, a ballad is found to deal with some event of mediaeval Portuguese history, the learned touch is immediately discernible, and the folk-quality noticeably absent, as in the ballad of *Inez de Castro*:

> From the rich palaces of Coimbra fared forth the noble Prince,
> With his pages and servants in royal cavalcade;
> From the beauteous Lady Inez with love his leave he took.
> Little knew her spouse that ne'er again would he behold her.
> "Whither go you, my Lord Prince? Ill betide this cavalcade:
> An evil fate, my Lord Dom Pedro, will befall this faring forth.
> Hasten home to your palaces, for they are slaying your delight."
> Left by herself the wife is exposed to evil deeds.
> The whole night through her tender voice was heard,
> Singing her sorrows to a melancholy air.
> "My Prince, my Lord, who gavest me thy royal hand,
> Listen where'er thou art, to the song of thy Inez.
> No longer can my sighs attain thy heart;
> Let mounts and valleys echo the song of thy Inez.
> This house is filled with mournful lamentations
> And in this palace nought is heard save thy Inez' song."

What the folk would seem above all to have demanded of the wandering minstrels from whom undoubtedly they first learned these songs, were *historias de pasmar*, marvellous tales as remote as possible from the humdrum round of their lives. Hence these legends, dramatic or pathetic, of mythical counts

[1] Pedro Fernandes Tomás, *Velhos Romances e Canções Portuguêses*.

and countesses, of kings of France or Aragon, of emperors of Germany and exalted (if rather vague) personages such as John of Gibraltar. The fact is that these tales, which they did not always bother to transpose into a Portuguese setting or to set up with Portuguese dramatis personae, are mostly versified versions of the common stock of European legend.

Of the four main groups into which the ballad literature of Europe is divided, the Portuguese *rimances* belong to the Latin cycle which they share with France, Catalonia, Northern Italy and Castile (excepting the great epic cycles of the last-named). The ballad *genre* made its first appearance almost simultaneously in the East and West during the twelfth century. At first considered too popular and frivolous to be written down, its transmission was exclusively oral until the fifteenth century, from which the earliest surviving examples date. During this literary "Dark Age" of some three hundred years the ballad themes migrated from end to end of Europe with singers the range of whose wanderings was far greater than is often realised. Playing Sancho Panza to the Quixote of learning, the ballad often followed the more important literary currents such as that which during this very period carried the troubadours from Provence and Burgundy to Portugal.

As to the themes, ballad-mongers, like Shakespeare, took theirs where they found them. Those of the Portuguese *rimances* are, for the most part, fairy-tales or (in the case of the *xácaras*) mediaeval "merry tales", ultimately of Eastern origin which they betray in their taste for splendour and for fixed formulae such as the use of the numbers three and seven. The Italian *novelle* formed the most probable transition between these Eastern tales and the Portuguese ballads. One at least of the latter, *Rainha e Captiva*, is derived from the Byzantine tale of *Flos and Blankflos* which was probably brought back from the East by the Crusaders. This story is mentioned by several poets, among them King Dinis, of the thirteenth and fourteenth centuries, and still enjoys popular currency. Aubrey Bell mentions a chapbook edition, published at Oporto in 1912, of the *Historia de Flores e Branca-Flor, seus amores e perigos que passaram por Flores ser mouro e Branca-Flor christã*. Incidentally the prose versions of this tale well reveal how faultily and incompletely the ballad-poets reproduced their original material.

The fact that such a legend could still enjoy any measure of popularity in the twentieth century should not be allowed to foster any illusions that the *rimance* is not a thing of the past. Of all Portuguese popular art-forms it is the only one which must be regarded as dead beyond recall, displaced by more garish forms of amusement.

CHAPTER X

THE POPULAR QUATRAIN

Cantigas são criancices,	Songs are childishness,
Palavras leva-as o vento;	Words are carried away by the wind;
Quem se fiar em cantigas	Who places his trust in verses
É falho de entendimento.	Is lacking in wisdom.

Popular poetry has peculiar charms and qualities which may be sought in vain in cultivated verse. Comparison between the two is fruitless. If the popular muse has well-defined limitations, she has moments of ineffable freshness of a kind seldom attained by her more sophisticated sister. Sincerity, directness, concision, spontaneity and human appeal are the qualities to which folk-poetry can lay claim, and there is no literature in which they are more prominent than the Portuguese.

The ballad and the detached lyrical quatrain are the two most characteristic folk-forms of the Iberian Peninsula. Since the spirit of Portugal is primarily lyric while that of Spain is dramatic, it is easily understood that the former country, while yielding the palm for ballads, can claim to retain it where the quatrain is concerned.

Too brief for the unfolding of a dramatic plot, the form has just that concision which can save lyrical sentiment from becoming mawkish. Even the commonplace verse you may hear your maidservant sing at her work is redeemed from cheapness by its brevity:

Ai de mim que triste sorte   Ah me, how sad a fate
É esta de amar,   Is this of loving,
Quando não se tem esp'rança   When there remains no hope
E só se pode suspirar.   And nothing left but sighs.

Such quatrains are innumerable. Upwards of sixty thousand have already appeared in print, of which one collector, Senhor Pires of Elvas, is responsible for ten thousand. They are comparable only to the Japanese three-line *hokku*. Inspiration comes so easily to a nation of poets, that verses spring up like the wild flowers of the transient Portuguese spring, and unless they are pressed within the leaves of a book are as soon forgotten. As pollen is carried by the honey-seeking bee, so are they carried from one end of Portugal to the other by military recruits or by those bands of labourers who bring in the harvest in the Alentejo and the vintage on the Douro. A few have had the good fortune to be enshrined in anthologies like Agostinho de Campos' *Mil Trovas* or to achieve fame through the appreciation of some famous literary figure. One such *quadra* aroused the enthusiasm of Byron and Alfred de Musset:

Chamas me tua vida,   You call me your life,
Eu tua alma quero ser.   But I would be your soul.
A vida acaba com a morte,   Life ends in death,
A alma não pode morrer.   The soul can never die.

The formula is an easy one. The metre is the same as that of the ballad, i.e. the *redondilha maior*. Only the second and fourth lines need rhyme, and these, at a pinch, may be assonant. It offers an ideal form for spontaneous improvisation, and one of which these peasant poets avail themselves to the full. I have even heard a labourer of Aldeia Nova de Ficalho express in verse his satisfaction at singing before a foreigner. The people of the Minho have a special gift of improvisation, which they exercise in *descantes* or *cantigas ao desafio* (poetical contests), often between a man and a woman, in which verse follows verse until one of the two competitors is discomfited. The following is a frequently quoted example of peasant repartee:

Todos quantos aqui estão   All you who are present
Sejam boas testemunhas   Bear faithful witness
De que esta mulher é minha That this woman is mine
Da cabeça até as unhas.   From her head to her toenails.

A minha alma é de Deus;	My soul belongs to God,
Minha cabeça é do rei;	My head to the King,
O corpo, da sepultura;	My body to the tomb,
As unhas, eu t'as darei.	But my toenails are yours for the taking.

These songs are closely interwoven with the life of the people. They are associated with all their activities and occasions. Strung together like beads, they may be heard at all the gatherings where the peasants assemble to share each other's work; at the *vessada* (ploughing), *desfolhada* (maize-stripping) and *espadelada* (flax-braking), all of which are made the occasion for merrymaking far into the night. They spring to the lips of the Douro vintagers as they pluck the purple grapes or tread them in the ice-cold *lagares*, of the reapers of Serpa when in the early August dawns they go out to the harvesting, of the gaily clad girls of Viana on their way to some rustic *romaria* or the mountain maids of Estrela as they turn in the interminable round dances. The attentive stranger comes to feel in them the perfectly appropriate *leit-motiv* of mountain and heath, of pine-encircled homestead and lonely dune-fringed shore.

A number of quatrains associated with some special occasion or tradition have already been quoted in this book. But the large majority of them are more general in theme. As might be expected, they are concerned before all else with love, and, by reason of the wistful, melancholy outlook of the race, with the sorrow of love rather than with the radiant happiness of reciprocated affection:

Quem ama sem ser amado,	He who loves but is not loved,
Que triste vida não tem:	How sad a life is his:
É ter pão, morrendo á fome,	'Tis having bread yet going hungry,
É ser filho e não ter mãe.	'Tis being a son without a mother.

But then song, in the view of these peasant poets, is of its very nature sad.

O canto é para os tristes	Song is for the sorrowful,
Quem o pode duvidar?	Who can doubt it?
Quantas vezes cantarei	How many times have I not sung
Com vontade de chorar.	When I would have wept.

There is indeed no sorrow deep enough to satisfy the unhappy lover:

Tudo que é triste no mundo	All the sorrow in the world
Gostava que fosse meu;	I would that it were mine,
Para ver se tudo junto	That I might see if gathered together
Era mais triste que eu.	It were more sorrowful than I.

Luxuriating in his grief, he feels that anything bright or cheerful is unbecoming to him:

O' flores do meu jardim,	O flowers in my garden,
Secai vos que mando eu;	Wither at my command;
E bom que não tenha flores	It is not right he should have flowers
Quem o seu amor perdeu.	Who has lost his love.

In the romantic notion of love there is not only sadness, but also an exaltation which is conveyed with passionate energy:

Se me encontrares cadaver	If you find me a lifeless corpse
Á porta duma ermida,	At the door of some wayside shrine,
Nem sequer c'o pé me toques	Touch me not with so much as your foot
Que posso voltar á vida.	Lest I might return to life.

Nor must it be thought that these poems express only an egotistical and self-pitying love. What could be a more touching expression of self-abnegation than this little verse:

Não choro por me deixares	I weep not because you have left me
Que o jardim mais flores tem:	For the garden has many more flowers:
Choro por não encontrares	I weep because you will never meet
Quem te queira tanto bem.	One who will love you so well.

Felicitous comparisons are found for both the happy and the unhappy lover:

Malva verde que se enleia	The green mallow winds itself,
Que se enleia pelo trigo;	Winds itself about the corn,
Quem me dera ser enleio	Oh that I too might be a mallow
Que eu me enleara comtigo.	And wind myself about you.
Coitadinho de quem nasce	Wretched is he who is born
No mundo sem ter ventura:	Into the world without luck:
E como o prato que quebra	He is like the plate that is broken
Que atiram com ele á rua.	And thrown out in the street.

The mixed feelings which love inspires are prettily described:

Quero ver-te e não te ver,	I want to see and yet not to see you,
Quero amar-te e não te amar,	I want to love and yet not to love you.
Quero-me encontrar comtigo,	I want to find myself with you,
Mas não te quero encontrar.	Yet not to go to meet you.

To these rustic lovers, the heart is something concrete, as solid and tangible as a gingerbread heart at a fair:

Aqui tens meu coração	Here you have my heart
A chave para o abrir;	And the key to unlock it;
Não tenho mais que te dar,	I have no more to give you,
Nem tu mais que pedir.	And you no more to ask.

To those who have been accustomed to regard peasants as materially minded folk seeing in Nature only the hard taskmaster from whom a livelihood must be wrung by an unremitting struggle, the sense of beauty shown in these poems will be a revelation:

Já lá vai o sol abaixo,	There sinks the sun to rest,
O sol vai, a sombra fica;	The sun sinks, the shadow stays;
Vai o sol admirado	The sun departs in wonder
Da sombra ficar tão rica.	At the richness of the shade.

Unadulterated nature-poems are rare. The observation of Nature is more usually wedded to some abstract conception, and the quatrain is divided into two halves, the first painting a concrete picture and the second expressing a thought or emotion. Frequently a clear symbolical connection may be discerned between the two parts:

O' rôla que vais voando	O dove, swiftly flying,
A fugir do gavião:	Before the falcon's claws:
Tambem a fugir do amor	Even so to escape from love
Tem ido o meu coração.	My heart has fled away.

A cana verde do mar	The green weeds of the sea
Deita raizes na areia;	Thrust roots into the sand;
Sou leal a todo o mundo	I am faithful to all the world
Todo o mundo me falseia.	And all the world betrays me.

In some instances this symbolism is more apparent than real:

Não há cousa que mais cheire	There is nought that smells sweeter
Do que a flor da alfazema;	Than the flower of lavender;
Não há gosto neste mundo	There is no pleasure in this world
Que não venha dar em pena.	Which does not end in pain.

In many verses, indeed, there is no clear parallel between the two halves, only at the most a faintly discernible affinity of mood, leaving more to the imagination, and imbuing the poem with a peculiarly delicate quality:

Lá no monte já cae neve,	Now snow is falling on the hills,
Caiu a flor ao sargaço;	The lily has lost its flower;
Não faças conta commigo	Take no heed of me
Que eu de ti conta não faço.	For I take none of you.

A oliveira da serra	The olive tree on the mountain
Do vento é combatida;	Is battered by the wind;
Inda espero de acabar	My hope is ever still to spend
Comtigo a minha vida.	The rest of my days with you.

Just as nature-pictures are used to illustrate human thoughts and emotions, so do the folk use birds or flowers to represent human actors:

O ladrão do negro melro	The rascally, thieving blackbird
Toda a noite assobiou;	Sang the whole night through;
Pela fresca madrugada	In the fresh hour of dawn
Bateu as asas, voou.	He took to his wings and flew.

The rose and carnation figure in countless verses. The motive is doubtless of Eastern origin, but in the eyes of the peasants the rose clearly symbolises a maiden and the carnation her lover:

Se quereis, rosa, ser rosa,	If, rose, you would be a rose,
Fugí do cravo, fugí!	Fly from the carnation, fly!
No tempo em que eu era rosa	In the days when I was a rose
Por um cravo me perdi.	Through a carnation I was lost.

O' rosa tu não consintas	Never consent, O rose,
Que o cravo te ponha a mão,	That the carnation lay hands on you,
Uma rosa enxovalhada	For a rose once sullied
Ja não tem aceitação.	Is rejected by all.

Many different flowers and fruits are used figuratively to designate a sweetheart. Rosemary, marjoram, yellow-tinted pear, green bramble, little yellow reed, garden of cannas, red lily in the undergrowth, are only a few of the terms in which the beloved is addressed. The blackberry is a more unexpected term of endearment until it is realised that it derives from the pun on *amora* (blackberry) and *amor* (love). At least, the Portuguese do not, like the Serbs, address the maidens of their choice as "unripe figs". But their comparisons are not invariably flattering:

Minha maçã vermelhinha	My little rosy apple
Picada do rouxinol;	Pecked by the nightingale;
Quem te picou que te coma	He who pecked you let him eat you
Que te picou no melhor.	For he pecked the best of you.

Of the examples quoted up to this point the greater part have been amorous in subject and elegiac in feeling. The last introduces a group in which a sly, satirical peasant humour contrives to peep through. Women, doctors, and priests are, in Portugal as elsewhere, the most usual victims of its shafts:

Eu amava-te, menina	I would love you, maiden,
Se não fosse um senão,	Were it not for one "but";
Seres pia d'água benta,	You are a stoup of holy water
Onde todos poem a mão.	In which all men dip their hands.
Andam as mulheres	There go the women
Enganando o mundo;	Cheating the world,
Com aneis de prata	With rings of silver
E elas são de chumbo.	And they but of lead.
Há duas cousas no mundo	Two things in this world
Que eu não posso comprender	I cannot conceive:
Um padre não se salvar	A priest being damned
E um cirurgião morrer.	And a doctor dying.
De Lisboa me mandaram	From Lisbon they sent me
Quatro frades num seirão;	Four monks in a basket;
Frei Azeite, Frei Vinagre,	Friar Oil, Friar Vinegar,
Frei Alho e Frei Pimentão.	Friar Garlic and Friar Pepper.

What canvas could portray the four monks more vividly than the two final lines of this last quatrain?

If peasant humour finds expression in these verses, so also do peasant wisdom and piety. The reader will find many wise counsels among them:

Ser pobre e casar com pobre	To be poor and marry a pauper
E remar contra a maré;	Is like rowing against the tide;
Casar com mulher sem dote	To marry a dowerless wife
E andar só com um pé.	Is like walking on only one foot.

The religious poems are often a moving expression of faith and devotion:

Meu menino Jesus	O my Infant Jesus
Descalcinho pelo chão,	Walking barefoot on the ground,
Metei os vossos pezinhos	Set your little feet
Dentro do meu coração.	Within my heart.

O pouco que Deus nos deu	For the little that God has given us
Cabe numa mão fechada;	There is room in one clenched fist;
O pouco com Deus é muito,	But that little with God is much,
O muito sem Deus é nada.	And much without God is nothing.

No ventre da Virgem Mãe	In the womb of the Virgin Mother
Encarnou Divina Graça;	The grace of God was made man;
Entrou e saiu por ela	It entered and passed through her
Como o sol pela vidraça.	As sunlight passes through glass.

Much experience and philosophy are here distilled:

Não há Sabbado sem sol,	No Saturday without sunshine,
Nem rosmaninho sem flor,	No lavender without flower,
Nem casada sem ciume,	No married wife without jealousy,
Nem solteira sem amor.	No single maid without love.

A Primavera tem lindas flores,	Spring comes in with flowers,
São bonitas mas não são iguais;	They are pretty but never the same;
A Primavera vai e volta sempre,	Spring passes, and always returns,
A mocidade não volta mais.	Only youth returns no more.

But what use has Youth for philosophy?

O livro da experiencia	The book of life's experience
Nenhum fruto ao homen dá;	Brings no profit to man;
Tem o conceito no fim	The moral is at the end
Ninguem o lê até lá.	And no one reads it so far.

The whole life and culture of the Portuguese peasants are concentrated in these verses. A few of them, encroaching on the preserves of the ballad, tell a story with that indirect, allusive style, and that gift for condensation and the suppression of the inessential, which are the hall-mark of all folk-art. The dumb, relentless tragedies of a lifetime are compressed within four lines:

Eu caí em me casar,	I slipped into marriage,
Dei o ouro pelo cobre;	I changed my gold for copper;
Dei a minha mocidade	I bartered my youth
Por dinheiro que não corre.	For a coin which has no tender.

Others are vivid *genre* pictures or miniature scenes of country life. The following example is in the dialect of Miranda do Douro:

Bi benir la gaita	I saw the pipes come
Al cimo del lhugar;	To the top of the village;
Pousei la mie roca	I laid aside my distaff
I pus-ma bailar.	And sprang up to dance.

Quatrains relating to different customs and superstitions are sprinkled throughout this book, but some as yet unmentioned may not come amiss here. The belief that red objects are endowed with a special virtue is reflected in a love-poem in which the roof and its tiles are doubtless symbolical of the maiden's favours:

As telhas do teu telhado	The tiles of your roof
São vermelhas, têm virtude;	Are red and full of virtue;
Passei por elas doente,	Being ill, I passed by them,
Logo me deram saude.	Forthwith they restored me to health.

In another poem the use of jet as an amulet is linked with a curious belief that a ring which "jumps" about on the finger is an evil omen:

Anel d'azeviche preto	My ring of black jet
Anda-me aos saltos no dedo;	Is jumping about on my finger;
Eu ando ameaçado	I am being threatened
De quem tenho pouco medo.	By one whom I fear but little.

Local pride and nostalgia have given rise to many verses in praise of the different regions and localities of Portugal:

Lisboa, com ser Lisboa,	Lisbon, for all it be Lisbon,
E ter navios no mar,	And the ships that sail on the sea,
Não é como a minha terra,	Is not, as my own home is,
A mais linda em Portugal.	The fairest in Portugal.
Mirandela, Mirandela,	Mirandela, Mirandela,
Mira-a bem, ficarás nela;	Look at it well and you will stay there;
Quem Mirandela mirou	Who once looked at Mirandela
Em Mirandela ficou.	Left Mirandela nevermore.

Strangely enough, although we looked long and closely at Mirandela from the many arched bridge over the Tua, we gave the poem the lie and stayed there no longer than was necessary for lunch.

The vast plains of the South, and the slow-moving, broad-hatted shepherds who wander over their face, are admirably rendered in another verse:

Alentejo não tem sombra	Alentejo has no shade
Senão a que vem do ceu.	Save that which falls from above.
Abrigue se aqui menina	Take shelter here, O maiden,
Debaixo do meu chapeu.	Beneath the brim of my hat.

Loyalty seems to require that to glorify one's own home, one must decry one's neighbour's. Accordingly, these topographical quatrains are disparaging as often as flattering:

Se fores a Landim	If you go to Landim
Leva contas para rezar,	Take your beads that you may pray,
Que lá é o Purgatorio	For it is the very Purgatory
Onde as almas vão penar.	Where departed souls find torment.

The reputation of many a town or village depends on that of its women, and these are specially singled out for praise or reprobation:

As meninas de Figueira	The maids of Figueira
São finas como o arame;	Are as sharp as wire;
Não ha dinheiro que as pague,	No money can buy them,
Nem rapaz que as engane.	Nor gallant deceive them.

Raparigas de Buarcos	The girls of Buarcos
São feias mas cantam bem;	Sing well but are not comely;
Quando vão a abrir a boca	When they open their mouths to sing
Cabe lhe um pão de vintem.	There is room enough for a penny loaf.

If local jealousies are expressed in these verses, it is only natural that they should also reflect the distaste with which every race seems inevitably to view its neighbours:

O sol nasce de Castela,	The sun rises in Castile,
Queres meu bem que nos lá vamos?	Shall we go thither, my love?
Não gosto que o sol esteja	I do not like the sun to be
Em poder dos Castelhanos.	In the hands of the Castilians.

Se houver de tomar amores	If I am to fall in love
Hei-de escolher um galego;	I shall choose a Galician;
Se me der a fome em maio;	Then if hunger comes in May;
Arre burro! Vou vendê-lo.	Gee up donkey! I'll go and sell him.

It is instructive, in examining these verses, to note the simple means of expression on which they rely. The effect of many of them depends on the use of apt and original images:

Como o vento é para o fogo	As wind is to fire
É a ausencia para o amor;	So is absence to love;
Se é pequeno, apaga-o logo;	If, slight, it puts it out;
Se é grande torna-o maior.	If great, it fans it to flame.
Minha mãe, case-me cedo	My mother, marry me early
Emquanto sou rapariga;	While I am still a girl;
Que o milho sachado tarde	For the crop that is weeded late
Não dá palha, nem espiga.	Gives neither ear nor straw.

There are fanciful and happy figures to express eternity:

Quando o sol deixar de dar	When the sun no longer strikes
Na ponta do alto freixo,	On the crest of the old ash tree,
Então saberás menina	Then, maiden, will you know
A razão porque te eu deixo.	The reason why I leave you.

and impossibility:

Nunca vi figueira alguma	I never saw a fig tree
Dar os figos na raiz;	Bring forth figs at the roots;
Nunca vi rapaz solteiro	I never saw a bachelor lad
Ser constante no que diz.	Be faithful to his word.

Principally, however, it is their great concision that makes the best of these verses so memorable. Examples throng to the mind:

Triste sorte é o de nascer,	It is an ill fate to be born,
Depois de nascer, pecar,	And being born, to sin,
Depois de pecar, morrer,	And having sinned, to die,
Depois de morrer, penar.	And being dead, to pass through torment.
Eu sou sol e tu es lua;	I am the sun and you are the moon;
Qual de nós será mais firme?	Which of us will be more constant?
Eu como sol a buscar-te,	I, the sun, in pursuing you,
Tu como sombra a fugir-me?	Or you, the moon in fleeing me?
Não ha sol como o de maio,	No sun like that of May,
Luar como o de janeiro;	No moonlight like that of January,
Nem cravo como o regado,	Nor carnation like that which is watered,
Nem amor como o primeiro.	Nor love like the first love of all.

Plate XV

A *FADO* IN THE STREET

[*facing p.* 240

Plate XVI

COIMBRA

Da terra sae a videira,	The vine grows out of the earth,
Saem da videira as uvas;	And out of the vine, the grape;
As solteiras são casadas'	Wives grow out of maidens
E as casadas são viuvas.	And widows out of wives.

Por te amar perdi a Deus;	To love you I lost my God;
Por teu amor me perdi,	Through your love myself was lost;
Agora vejo me só,	And now, behold me alone,
Sem Deus, sem amor, sem ti.	Without God, without love, without you.

This poem might be described as a four-line epitome of Camilo Castelo Branco's novel *Amor de Perdição*, a tragedy of all-consuming love which is perhaps the most famous and certainly the most popular work of nineteenth-century fiction in Portugal.

The origin and growth of the popular quatrain are difficult to trace with any certainty. A dance-song before all else, it is in this respect the heir of the *cossante*, with which it shares the use of a refrain and an occasional tendency to repetition in *leixapren*.

The refrain is most usually a meaningless or irrelevant phrase such as *O ai o linda* or *Meu lirio roxo*. It belongs to one particular air (*moda*) to which quatrains are sung, rather than to any special verse. It may be introduced in the middle of the verse or at the end, or after any or all of the lines:

Á oliveira da serra,	The olive tree in the mountains,
O vento leva a flor,	The wind carries away its flower,
O ai ó linda	*O ai ó linda*
Só a mim ninguem me leva	There is none to carry me
O ai ó linda	*O ai ó linda*
Para o pé de meu amor.	To the feet of my love.

It is not unusual for a whole quatrain associated with a particular tune to serve as a refrain (*resquebo*), and to be sung once through in chorus between each solo verse.

The influence of *leixapren* is seen in pairs of quatrains, in which the second is a rearrangement of the first with a line or two altered:

O' pavão, lindo pavão,	O peacock, pretty peacock,
Lindas penas o pavão tem,	Fair feathers has the peacock;
Não vi olhos para amar	Never saw I eyes for loving
Como são os de meu bem.	Like those of my darling.

Como são os de meu bem,	Like those of my darling,
Como são os de minha amada.	Like those of my beloved.
O' pavão, lindo pavão,	O peacock, pretty peacock,
O' pavão, pena riscada.	O peacock, with patterned feathers.

Neither in the *Cancioneiros*, nor in Gil Vicente's plays, where popular motives abound, are independent quatrains to be found. This and the fact that their metre is that of the *rimance* suggest that they are an offshoot of the latter. Indeed, some of the quatrains are fragments taken bodily out of older ballads. It would be useless therefore to seek the origin of the form farther back than the sixteenth century at the earliest. The lack of internal evidence leaves ample room for conjecture.

In any case these verses have moved with the times, and have been modified by such innovations as gas-lighting and railways:

Na estação da Vila Velha	In the station of Vila Velha
O maquinista chorava,	The engine-driver was weeping,
Acabou se lhe o *cravão*	He'd run out of coal
O comboio não andava.	And the train wouldn't go.

If the quatrains are difficult to date, it is no easier to estimate what proportion of the credit for their invention is really due to the peasants themselves. That many of them are improvised by the folk cannot be denied. But peasant improvisation does not necessarily imply original invention. It often amounts to no more than the refashioning of materials which the folk have generally borrowed in the first place and thereafter preserved to form a gradually built-up tradition.

It will not have escaped the attentive reader that many of the verses quoted in this chapter bear the imprint of a cultivated, literary inspiration. Some of them could never have been composed by peasants. Others reflect the mannerisms of lettered poets, simplified and popularised by long years of oral transmission. In a few instances, a definite connection may be established between a quatrain current among the people and a printed poem by a known writer.

O' amor vai e vem logo	Não passeis vos, cavaleiro,
Volta depois por aqui,	Tantas vezes por aqui,
Que eu abaixarei meus olhos	Que eu abaixarei meus olhos
Jurarei que te não vi.	Jurarei que vos não vi.

Of the two verses printed above, the first is a popular quatrain recently collected and published in the *Revista Lusitana*. The second, of which it is clearly a corruption, is taken from a poem by a sixteenth-century poet, Cristovão Falcão.

Coimbra must have played a larger part than is generally admitted in the creation of these verses. The quantity of verse produced by the University has always been considerable and reflects its consciously romantic tradition and atmosphere. The songs of student poets doubtless find a ready echo on the lips of the pretty *tricanas* of the Mondego valley. Moreover, the students themselves, many of whom are of humble origin and come from all parts of the country, disseminate them on their return to their homes. Thus both as poets and distributors they form an important link between the erudite and the popular muse. Many of those who are still alive to-day have seen the lines in which they enshrined their youthful indiscretions adopted anonymously by the folk, and thus become popular in the widest acceptance of the word.

As in the music to which they are sung, there runs through these verses a vein of graceful artificiality which corresponds with the rustic rococo of the rural architecture. Without being pretentious or ponderous, they are pleasantly stylised, making light play with the well-worn themes of bucolic poets, delighting in concetti, quips and simple puns.

What could be prettier, yet more artificial than the fancy expressed in the following lines:

Fechei na mão um sorriso	I closed my hand on a smile
Da tua boca formosa;	From your lovely lips;
Quando fui abrir a mão	When I opened my hand
Vi-a toda côr de rosa.	I found it tinted with rose.

or the play on the word *pena* (sorrow or pen) which in one form or another recurs in countless quatrains:

Com pena peguei na pena,	With sorrow I took up my pen,
Com pena pus me a escrever;	With the pen I set me to write;
Caiu me a pena da mão	But the pen slipped out of my hand
Com pena de te não ver.	With sorrow that I could not see you.

The lyrical quatrain thus reflects the intimate connection which in Portugal more even than in most countries, links the popular and the cultivated arts. The contribution of the folk to their own poetry amounts less to active invention than to a gradual process of assimilation and transformation, capable on occasion of lapsing into deformation.

In the main, however, the influence of the folk is beneficial. Just as running water polishes a stone, removing its asperities, rounding and perfecting its form, so does the current of oral transmission remove from these verses all individual touches foreign to the instincts of the mass, and bring out all that is most universal and most faithful to the national outlook. Rarely does any final and definite form emerge from the gradual metamorphosis to which the quatrains are subjected, but once they have achieved currency as anonymous folk-poems, they have come already to express the genius of the race as fully as though they were in actual fact its original communal creation.

CHAPTER XI

THE FADO

"The *fado*, the knife and the guitar," writes Pinto de Carvalho, "are the three favourites adored by the people of Lisbon." [1] If the foreign visitor to Portugal's capital is likely to be disappointed in any hopes he may entertain of seeing the second member of this trio at work, he will find the other two more easily accessible.

Although there are several well-known *boîtes de nuit* where the *fado* is sung to an audience drawn principally from the *demi-monde*, it is better heard in simpler, more plebeian surroundings. Its true home is Alfama and Mouraria, the poor quarters of the city, which flaunt their picturesque squalour on the slopes below St George's Castle. A walk through these steep, narrow streets on a moonlit night is likely to be rewarded with the sound of a guitar and the mournful cadences of the *triste canção do sul*. But to hold it surely in one's grasp it is best to go to one of the popular cafés such as the "Luso" and the "Victoria" where it is regularly performed by semi-professional *fadistas*.

The social standing of these places seems to be largely a matter of headgear. Entrance to the first is forbidden to those

[1] *Historia do Fado.* Lisbon, 1903.

wearing caps or bérets (a fine distinction). On the other hand, the patrons of the second, mostly seafaring folk, seem to wear no other head-covering. The spacious rectangle of the "Luso" and the low-vaulted room of the "Victoria" are alike crowded with tables and chairs at which many men (but few women) sit drinking coffee, beer, or soft drinks with exotic names such as Maracuja or Guarana. Presently the lights are lowered and turn red, and a woman steps on to a low platform at one side of the room. She is the singer. Her accompanists, seated in front of her, are armed with guitars of two different types, known the one as a *guitarra* and the other as a *viola da França*. The Portuguese apply the name of *guitarra* to an instrument with a rounded soundboard and six double strings of wire. The *viola da França*, on the other hand, is but another name for the familiar Spanish guitar. The tune of the *fado*, or figured variations upon it, is played on the former instrument which has a sweeter, more silvery tone, while the latter is used to provide a thrumming accompaniment, alternating invariably between the chords of the tonic and the dominant seventh.

After a few bars of this accompaniment the *fadista* begins her song. With head thrown back, eyes half closed, ecstatic expression and body swaying slightly to the rhythm of the music, she sings in the curiously rough, untrained voice and simple, unpretentious manner which are dictated by tradition. The *fado* does not lend itself to *bel canto*, and the opera singer with his cultivated voice and professional manner would never be tolerated by the critical audience that listens in unbroken silence to these songs. Against the strict common time of her accompaniment the *fadista* maintains a rhythm as free and flexible as that of the jazz-singer, with whom she shares certain tricks of syncopation and anticipation or suspension of the rhythmic beat which give the song a lilt as fascinating as it is difficult to reproduce.

Leaving aside all technical considerations, it is difficult to put into words the very individual character of the *fado*. Epithets crowd into the mind, all appropriate up to a point, yet many of them mutually contradictory, and none of them conveying the exact and complete impression that one strives in vain to seize and to evoke.

The true *fado* is always sad. Usually in the minor, it retains even in the major the melancholy character nowadays associated with the minor. It may be wild, finely exultant in its sadness, seeming to revel in tragedy; or, more often, striking a note of pathetic and almost languid resignation. Its sophisticated cadences breathe a spirit of theatrical self-pity combined with genuine sincerity. It is emotional, passionate, erotic, sensuous, one might say meretricious, and yet, like some rustic courtesan, fundamentally simple and unpretentious. Perhaps this is because these qualities, however irreconcilable to the Anglo-Saxon way of thinking, are nevertheless reconciled in the Portuguese temperament, or at least in one aspect of it.

"Both words and music", writes Pinto de Carvalho, "reflect the abrupt turns of fickle Fortune, the evil destiny of the unfortunate, the irony of fate, the piercing pangs of love, the poignancy of absence or despair, the profound sobs of discouragement, the sorrows of *saudade*, the caprices of the heart, and those ineffable moments when the souls of lovers descend to their lips and, before flying back on high, hover for an instant in a sweet embrace."

This analysis leaves little more to be said. And it is all the more surprising that the merits of the *fado* are as hotly disputed by some Portuguese writers as they are loudly proclaimed by others.

On the one hand, Ventura de Abrantes, writing in the *Guitarra de Portugal*, describes these as "the most Portuguese of all songs and the liturgy of the nation's soul". And the honour of having given them birth is disputed in the columns of the daily press by Lisbon and Coimbra, where black-gowned students stroll through the narrow streets on moonlight nights twanging their guitars and singing serenades to shadowed casement windows.

On the other hand there is no lack of those who are violent in their condemnation of what Alberto Pimentel calls "these deliquescent and immoral melodies".[1]

---

[1] *Triste canção do sul.* Lisbon, 1904.

"There is nothing in this order of things", writes A. Arroio, "which can be compared with the *fado* as an expression of the lowest types of melodrama and of the most exaggeratedly bad taste."[1] José Maciel Ribeiro Fortes goes so far as to stigmatise it as "a song of rogues, a hymn to crime, an ode to vice, an encouragement to moral depravity...an unhealthy emanation from the centres of corruption, from the infamous habitations of the scum of society".[2]

This virulent antagonism is perhaps to be explained in part by resentment at the manner in which foreigners, and indeed many Portuguese, have accepted the *fado* as the only popular musical expression of the Portuguese nation. It is a curious fact that, as Adolfo Salazar wrote some years ago in an article in *El Sol*, "that which is considered to be typical of a country, and which even the inhabitants themselves admit to be characteristic (*castizo*) is apt to be of the most recent importation and the least traditional thing in the world". Of the truth of this statement a hundred examples spring to the mind: the *zortzikos* of the Basques, for instance, the tzigane music of Hungary and many of the popular melodies which pass for Scotch or Irish. Its application to the *fado* is in the main justified. For the latter is very different from the true peasant folk-song of Portugal.

Salazar's words "of recent importation" are worth bearing in mind. The word *fado* is not used in Portuguese literature in the sense of a song until 1833 when the two terms *fado* and *fadista* make a simultaneous appearance in a broadsheet:

Dansamos tambem o Fado por ser dansa muito guapa	We dance the *fado* for the fine dance that it is
E tomamos um fadista que sabe jogar á faca.	And we take a *fadista* who knows how to use his knife.

Both words are undoubtedly derived from the Latin *fatum* (fate) and seem to have been first used in connection with the notorious characters of the worst quarters of Lisbon, who enjoyed a monopoly of the song until about 1870, when it began to acquire a wider popularity. It may be that the word *fado* was first applied to the songs in which these rogues lamented

[1] *O Canto Coral e a sua funcção social.*   [2] *O Fado.* Oporto, 1926.

their evil destiny. On the other hand, it is possible that their very destiny had already earned them the appellation of *fadistas*, and that the term was only subsequently extended to their songs. "The *fadista*", writes Pinto de Carvalho, "who played the rôle filled to-day by the Parisian *voyou* and the American 'rough'...was the product of all the vices and the incarnation of everything despicable." Fifty years ago, Lady Jackson noted that the *fadistas* "wear a peculiar kind of black cap, wide black trousers with close-fitting jacket, and their hair flowing low on the shoulders....They are held in very bad repute, being mostly *vauriens* of dissolute habits".[1] Their modern prototype is a Bohemian of a gentler order. Be that as it may, the problem whether the song took its name from the singer, or the latter from the song will probably never be solved. The more so since recent research has revealed that the term is found in Brazil as early as 1819. For this and for other reasons the possibility cannot be excluded that the name or the song or both came to Portugal from that continent, the whole of which, as Paul Morand declares in *Air Indien*, "weeps and regrets in music".

The question of the origin of the *fado*, like that of its merits, has long been the subject of learned, and correspondingly acrimonious, controversy. The brilliant but erratic Theophilo Braga claimed for it a Moorish origin,[2] and this theory finds rather surprising support in Edmundo Arménio Correia Lopes, who sees not only in the *fado* but also in the Spanish *tango* and *habañera* the direct descendants of the Arabian *majuri*.[3] Apart from internal evidence (for the *fado* is anything but Moorish in character) this theory is rendered inherently improbable by the fact that neither in Andalusia, nor in the Algarve, where the Moors remained longer than in other parts of the Iberian Peninsula, is the *fado* or any kindred form indigenous. Other writers have sought its birthplace on the high seas. "In my opinion", writes Pinto de Carvalho, "the *fado* is of maritime origin, an origin which is confirmed by its rhythm, undulating

---

[1] Catherine Charlotte Lady Jackson, *Fair Lusitania*. London, 1874, p. 106.
[2] *O Povo portuguez nos seus costumes, crenças e tradições.*
[3] *Cancioneirinho de Fozcoa.* Coimbra, 1926.

as the cadenced movements of the wave, regular as the heaving of a ship...or as the beating of waves upon the shore." Ventura de Abrantes combines this theory with that of a peasant origin: "The real creators of the *fado* were the sailors, soldiers and country folk. They it was who, on their journeys beyond the seas, on their high adventures of discovery, far from the land which gave them birth, and in order not to forget their great love for it, sang the *fado* to express their *saudades* and their longing to return." Professor Sampaio prefers to derive it from the innumerable variants of a single Portuguese folk-song, that to which in the Minho the peasants dance round the bonfires on St John's Eve. This theory cannot be accepted as it stands, although there is a grain of substance in it, as will be seen.

The hypothesis which finds most credence among musical circles in Portugal is one which has far more arguments in its favour. It is impossible to conceive the *fado* without a guitar accompaniment, and its origins therefore can hardly be sought in an environment where that instrument was not cultivated. "The *fado* and the guitar", writes Luis de Freitas Branco, "have no definite regional character in our countryside. They are characteristic of the populace of the great cities, and in particular of Lisbon."[1] The guitar is, of course, descended from an instrument introduced into the Peninsula by the Moors and adopted by the mediaeval minstrels under the name of *vihuela*. It formed one member of the trio (the others were tea and toast) said to have been introduced into England by Catherine of Braganza on her marriage to Charles II, and to have become so popular that the English were soon exporting guitars to Portugal, a circumstance which has led one Portuguese musical historian to the surprising conclusion that the English were the inventors of the Portuguese guitar.[2]

---

[1] *A Música em Portugal*. Lisbon, 1930.

[2] The Portuguese guitar closely resembles the cittern popular in England and Germany during the sixteenth and seventeenth centuries and stated by Vincentio Galilei (the father of the astronomer) to be of English origin. In the eighteenth century it was known as the English (as opposed to the Spanish) guitar. It will probably never be known whether Portugal influenced England in this respect or *vice versa*.

Since songs and dances to the accompaniment of the guitar
clearly existed in Lisbon long before the name of the *fado* was
heard, it seems reasonable to seek the immediate sources of
the latter in the popular music of Lisbon during the period
preceding its appearance. There are fortunately a number of
descriptions and references which afford a fairly clear picture
of the music of Lisbon during past centuries. Owing to its
geographical position and to its importance both as a port and
as a capital, the city has always been particularly receptive to
exotic influences and, as is generally admitted, to the admixture
of foreign, even non-European, blood. Since the great epoch
of discovery and colonisation during the first half of the sixteenth century the *lisboeta* has always shown a marked liking
for exotic song and dance, and in particular for those of the
negroid races with whom the Portuguese came into contact in
Africa, and whom they transplanted as slaves not only to Brazil,
but also to extensive regions in Southern Portugal which had
been left empty first by the Moors and later by the emigrants
to the New World.

It is possible to date almost exactly the period when musical
taste veered from the cheerful to the melancholy. In his
Triumpho do Inverno (1519) Gil Vicente places the following
lines on the lips of one of his characters:

Em Portugal vi eu já	In Portugal I once did see
Em cada casa pandeiro,	In every house a drum
E gaita em cada palheiro;	And pipes in every barn;
E de vinte annos aca	But now since twenty years ago
Não há hi gaita nem gaiteiro.	Neither pipes are there nor pipers.

"Jeremiah is our drummer, now", he continues, and the merry
songs of his youth have been replaced by others so doleful
"that they were evidently composed by some Jew of Aveiro
on the death of his grandfather." A hundred years later Duarte
Nunes de Leão mentions that it is the mark of the low-born to
prefer cheerful music to *hûa canção...de toada mui lamentavel*
("a song...in a most lamentable strain").

The literature of the last three hundred years is full of
references to such exotic dances. The earliest in point of time
is the *batuque* which seems to have become popular in Lisbon

in the sixteenth century. Others are the *oitavado* and the *arrepia*, both of which are mentioned in 1734, the *guineo* from Guinea, the *arromba* accompanied by the Spanish guitar and the slapping of thighs, the *zabel macau*, the *charamba*, the *sarembeque*, the *chegança*, the *canario* "danced with many difficult postures and great gravity", the *fofa* and *lundum*. Like so many native dances, most of these appear to have been rather lascivious and obscene in nature. It was doubtless their exotic character which appealed to the creole Joséphine Beauharnais, who, on seeing them danced in 1808 by Portuguese soldiers at Bayonne, is said to have exclaimed: "Oh, how I love these Portuguese gavottes." Under King Manuel I, a law was promulgated proscribing *batuques*, *charambas* and *lundums*, and the accounts of contemporary travellers leave little doubt of their character.

In 1761 the black slaves in Portugal were liberated, and many of them established themselves in the Alfama quarter of Lisbon, a fact which may have contributed to the extraordinary popularity enjoyed by the *fofa* towards the end of the eighteenth century. Dezoteux has left the following account of this dance: "The people ran about here and there singing and dancing the *fofa*, a sort of national dance performed in pairs to the accompaniment of the guitar or some other instrument; a dance so lascivious that decency blushes at witnessing it, and I would not dare to describe it."[1] Dalrymple, before whom it was danced by a coloured man and woman in 1774, calls it "the dance peculiar to Portugal as the *fandango* is to Spain... the most indecent thing I ever saw".

Like the *fado* in its older form, the *fofa* was thus danced, as well as sung, to a guitar accompaniment. But more important, from the point of view of the "fadographic problem the vastness and complexity of which would furnish abundant materials for two fat volumes",[2] is the *lundum* which, together with the *modinha*, shared the affections of the Lisbon populace from the last quarter of the eighteenth to the middle of the nineteenth

[1] *Voyage en Portugal.*   [2] Fortes, *op. cit.*

century. The *lundum* came to Portugal from Brazil. "It is of American origin", writes José Maria de Andrade Ferreira, "and often recalls the songs of Peru in its languidness and its gentle meanderings which reflect the indolence of countries devoured by the heat of a burning sun."[1] It seems certain, however, that it reached Brazil from the west coast of Africa. Alberto Pimentel writes of "the languid and monotonous song of the African native whence appears to have come the *lundum* which accorded with our innate melancholy and was certainly the popular song most closely related to the modern *fado*". The reference of the satirical poet Nicolau Tolentino de Almeida to the *doce lundum chorado* ("sweetly weeping *lundum*") coupled with an old print showing a "fadista dancing the *lundum*" suggest that this dance possessed the regular rhythm and the languorous, indolent lilt of the modern *fado*.

During the reigns of João V, José I and Maria I, that is to say throughout the middle section of the eighteenth century, "the people of Lisbon danced the *lundum* to the sound of the guitar" writes Carlos Malheiro Dias.[2] But it was not long before its character was modified by contact with the *modinha*, defined by Ernesto Vieira as "a sad and sentimental melody often in the minor" Of the *modinha* Beckford has left an unforgettable description:

Those who have never heard this original sort of music, must and will remain ignorant of the most bewitching melodies that ever existed since the days of the Sybarites. They consist of languid interrupted measures, as if the breath was gone with excess of rapture, and the soul panting to meet the kindred soul of some beloved object. With a childish carelessness they steal into the heart, before it has time to arm itself against their enervating influence; you fancy you are swallowing milk, and are admitting the poison of voluptuousness into the closest recesses of your existence. At least, such beings as feel the power of harmonious sounds are doing so; I won't answer for hard-eared, phlegmatic northern animals.

The two forms were equally popular in Lisbon, and Eduardo Noronha's assertion that the *fado* "must have had the *lundum*

---

[1] *Curso de literatura portuguêsa.* Lisbon, 1875.
[2] *Cartas de Lisboa.* Lisbon, 1905.

for father and the *modinha* for mother" is elaborated by Alberto
Pimentel: "The national plays performed in the Salitre and
Rua dos Condes Theatres contained Italian music, the most
catchy airs of which became public property and were transformed into the *modas* or *modinhas* which radiated all over the
country. There was even a periodical devoted to *modinhas*
which published the most interesting of them. In these pieces
were also interpolated *lundums*, African dances which served
as interludes. Gradually the *lundum* began to take on an
independent existence as a song which rapidly became the
favourite of the lowest grades of society who gave it the name
of *fado*."

That the dance of the *fado* was almost identical with that of
the *lundum* as performed in Portugal is suggested by a passage
from one of the *Improvisos* of Falmeno (Felisberto Inácio
Januário Cordeiro) who in a foot-note on the dances of Brazil,
writes that "*fado, tacorá*...are the names of Brazilian dances
corresponding to those which in Portugal are called *lundum,
fandango, fofa, chula*, etc."

There is nothing inherently improbable in the suggestion
that the *fado* is a sort of musical half-breed, heir to the rhythmic
features of one parent and the melodic features of the other.
The negro spirituals of the United States have been shown to
be the product of African rhythms and European hymn-tunes;
and the tango, maxixe and other dances of Latin America
combine the steady monotonous beat and indolent syncopation
of the coloured races with the formal symmetry and sophisticated line of European art-music.

The analogies which, in spite of wide differences, undoubtedly
exist between the *fado* and certain Portuguese country-songs
still remain to be accounted for, and to this end the internal
evidence offered by the *fado* itself must be taken into consideration. It may therefore be of value to quote a few examples
taken down either directly or through the medium of the
gramophone from authentic singers in Lisbon, Ex. 1–5, and
Coimbra, Ex. 6–8:

Ex. 1

Ex. 2

Ex. 3

Ex. 4

Andante

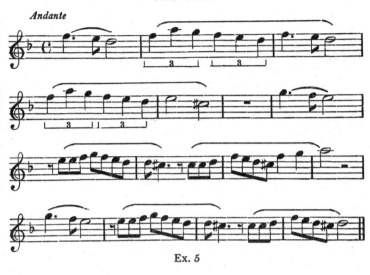

Ex. 5

Largo

Ex. 7

Ex. 8

With these examples in mind it is now possible to attempt a definition of the *fado*, excluding from it many types of song to which popular misconception has applied the name. The true *fado* is the combination of a rhythm, a type of melody, and a style, or rather two distinct styles, of singing, superimposed on a specific harmonic foundation.

According to Ernesto Vieira the *fado* consists of "sections of eight bars in 2/4 time, divided equally and symmetrically into two halves each consisting of two phrases". Apart from the fact that it is usually more convenient to transcribe the *fado* in 4/4 than in 2/4 time, this definition is accurate enough as far as it goes. But it could equally be applied to many of the rural dance-songs, and takes no account of the fundamental and characteristic rhythm of the phrase which is this:

It is this touch of syncopation, common to almost every *fado*, on which Dr Sampaio bases his claim, alluded to above, that the *fado* is derived from a group of songs sung by the Portuguese peasantry round the Midsummer bonfires. Here is one of these songs which I collected at Bombarral:

This is not, however, the only example of this rhythm found in the country music. Ex. 6 in chapter VIII is a case in point. There are several others in the Pedro Fernandes Tomás collection,[1] and it is significant that in each case the collector was informed by the singer that the song was of Brazilian

---

[1] *Velhas canções e romances populares portugueses.* Coimbra, 1913.

origin. It seems, therefore, that this feature is a Brazilian (i.e. originally African) contribution to the composition of the *fado*.

The regular beat of the rhythm, the formal symmetry of the structure, the falling cadences, the weak final accent—all these are common to both *fado* and country-song.

From the melodic point of view the most striking quality of the *fado* is its obviousness and its often banal facility. The instrumental accompaniment is limited to the chords of the tonic and dominant seventh (with an occasional modulation in the more modern examples to the same chords in the key of the dominant), and the melody is founded on, and limited by, this pre-existing harmonic scheme. Many of the rural dance-songs, as we have seen, are similarly founded on an harmonic basis, but they contrive usually to be more melodic in character than the *fados*, some of which are inconceivable, and indeed almost unsingable, apart from their accompaniment.

So far, therefore, the *fado* has been shown to be the product of a Portuguese rhythm, an exotic lilt derived from negro syncopation, and an elementary harmonic system. On this foundation the people of Lisbon sang tunes of the type congenial to them, some of which, perhaps, may have been inspired by peasant songs, but most of which must have been derived, to judge from their sophisticated cadences, from street music and the echoes of Italian opera.

This would account for the occasional analogies to be found between both branches of Portuguese folk-song, the urban and the rural, as also for the still greater difference of spirit which separates them, and which becomes still more marked in the manner of their singing.

Some account has already been given of the mannerisms of the Lisbon *fadista*. The most characteristic of these is the flexibility of the rhythm, a free *rubato* over the steady beat of the accompaniment, which it is extremely difficult to seize or to transcribe, and to which staff-notation imparts a rigidity the lack of which is its principal charm. The five notes unevenly distributed over the four beats of the bar stand to one another in countless slightly differing proportions of time value.

These subtle rhythmic inflections vary from verse to verse, and, like plain-song, though in a different way, the tune moulds itself to the plastic form of the words. So closely do these rhythmic mannerisms—together with the easy, intimate manner and the throaty, almost hoarse voice—resemble the style of the authentic jazz singer, that one is tempted to see in them a further negro legacy to the *fado*.

At Coimbra the *fado* has a very different character. Here it is no longer the song of the common people. It has become the property of the students who wander along the green banks of the Mondego, among the poplars of the *Choupal*, dreamily singing to their silvery guitars. Their clear, warm tenor voices give the song a character that is more refined, more sentimental, in a word more aristocratic. At Coimbra the *fado* seems to be divorced from everyday realities and cares, to be over-spiritualised, and to express a vague romantic yearning which is in keeping with the atmosphere of the ancient university city. It is the song of those who still retain and cherish their illusions, not of those who have  irretrievably lost them. But it is less individual and distinctive, more akin to the serenades of other parts of Southern Europe than is the *fado de Lisboa*. This impression is heightened in the stylised *fados* of António Menano, the merits of which are a bone of contention among Portuguese musicians.

It must not be thought, however, that the *fado*, being a fusion of elements not all of Portuguese origin, and being as much "popular" as "folk" in character, is aesthetically valueless. There are, of course, any number of so-called *fados* composed to be sung in revues or warbled in drawing-rooms by sentimental *meninas*. And, incidentally, it is exclusively these which find their way into print. The tunes to which the authentic *fados* are sung have never, to my knowledge, previously been taken down, although, curiously enough, there are excellent gramophone records of many of them. It is difficult to fix the age of these anonymous tunes; they are usually older than the words sung to them, since the *fadistas* have a great facility for improvising verses, which, however, they generally sing to old and well-worn tunes. But, even when they compose their own tunes, these are less the expression of the composer's individuality than the outcome of their environment and determined by a gradually-formed and well-defined tradition. They are the spontaneous music of the plebeian populace of Lisbon, just as the folk-songs are that of the peasants, and, like the *chansons vécues* of the French artisan, may perhaps most fairly be defined as "urban folk-song"

However little the *fado* may have in common with the folk-songs of the countryside, it is no less Portuguese in spirit than these latter, with which it shares the gift of expressing one of the most marked traits of Portuguese character. "There is indeed," writes Armando Leça, "a definite connection between the melting sadness of the *fado* and a certain amorous and sentimental fatalism fairly common in our race."[1] In Portuguese this quality is best expressed by the word *saudade* which has no exact equivalent in any other language. To a musician a well-sung *fado choradinho* will offer a better interpretation of this word than any verbal explanation, but it may profitably be supplemented by a passage from Aubrey Bell's *In Portugal*: "The famous *saudade* of the Portuguese is a vague and constant desire for something that does not and probably cannot exist, for something other than the present, a turning towards the past or towards the future; not an active discontent or poignant sadness, but an indolent dreaming wistfulness."[2] In a word *saudade* is yearning: yearning for something so indefinite as to be indefinable: an unrestrained indulgence in yearning. It is a blend of German *Sehnsucht*, French *nostalgie*, and something else besides. It couples the vague longing of the Celt for the unattainable with a Latin sense of reality which induces realisation that it is indeed unattainable, and with the resultant

[1] *Da Música portuguêsa*. Lisbon, 1922.
[2] *In Portugal*. London, 1912.

discouragement and resignation. All this is implied in the lilting measures of the *fado*, in its languid triplets and syncopations and, above all, in its descending and, as it were, drooping cadences.

That the *fado* is faithful to the national character or rather to one side of it, is shown by the way in which it has prospered when transplanted to the foreign soil of Oporto, and by the popularity which it has won in the countryside, where, like some fast-growing weed, it has choked and stifled the wild-flowers of peasant song.

If further proof be needed, it is afforded by the hold which the story of Severa, or rather her legend, has taken of the

popular imagination. Maria Severa, greatest of all *fadistas*, was the daughter of a woman known as *A Barbuda* (The Bearded Lady) who kept a tavern in Madragoa, the fish-wives' quarter of Lisbon. Somewhere about 1840, mother and daughter moved to the Mouraria and established themselves in the Rua do Capelão, which its unenviable reputation had earned the nickname of Rua Suja. From her mother, Severa learned to sing and dance (*bater*) the *fado*. Her moral upbringing went no further; and, like the Maid of Amsterdam, she soon became "mistress of her trade". Living in the atmosphere which engendered the *fado*, she grew to be its most renowned exponent. Her first lover, a man of her own class, was banished to Africa for a *crime passionnel*. Handsome, passionate, violent, in all things immoderate, she attracted the attention of a bull-fighting aristocrat with a taste for low life, the Conde de Vimioso (not, as Julio Dantas has it, the Conde de Marialva). The tempestuous love-affair between these two, of which the sentimental aspect has blinded the Portuguese to its more sordid side, is still sung in many *fados* and forms the foundation on which Julio Dantas has woven a story which, as play, novel and sound-film, has won widespread popularity. In this story Severa dies, like Mimi in *La Bohème*, of consumption. But popular tradition maintains more appropriately, if less romantically, that the real cause of her death, was a surfeit of pigeon and red wine.

No mention has yet been made of the words of these songs, and in truth they are of little value or interest. The poems improvised by the *fadistas* are composed in verses of four, five, eight or ten lines. A very usual form is that of the *glosa*, based on a quatrain, sometimes of popular origin, each line of which is taken in turn to form the last line of an eight-line verse. Full of exaggerated sentimentality, wallowing in self-pity, they have the poverty of expression and the hackneyed turn of phrase commonly associated with greeting cards. Many of them extol the joys of being unhappy and of expressing one's wretchedness through the medium of *fado* and guitar. Others relate long and doleful tales of blighted love, conjugal infidelity, horrible crime or undeserved misfortune. One that I heard was a sort of rhymed sermon on pacifism, ending up with an assertion by the singer that he had been inspired to compose it by seeing the film *All Quiet on the Western Front* at the São Luís Cinema. Another, exquisite in its bathos, described a bicycle race round Portugal, while a third, a lament for a footballer named Pepé (rhymed with crêpé), exclaimed:

Foi um astro, foi um sol,   He was a star, he was a sun,
Nós campos de futebol.   On the football field.

But if they achieve nothing else, these verses show that the *fado* is still alive. Despite their monotony and morbidity, the artless sophistication of the words and the technical poverty of the music, it is ridiculous to dismiss these songs, with

Pimentel, as "deliquescent and immoral melodies...to be understood and felt only by those who vegetate in the mire of crapulence". They have an attraction, difficult to analyse or explain, which may perhaps be accounted for, in part, by their peculiar blend of sincerity and sophistication, of freshness and conventionality; and they have a real intrinsic value as the expression of a racial mood and of a social environment.

CHAPTER XII

FOLK-TALES AND PROVERBS

Can European folk-lore offer anything more strangely baffling than the folk-tale? Those interminable stories told round the fire on winter nights, impressed by endless repetition on the minds of the young and by them handed down in later years to their children, might reasonably be expected to bear some relation to national history and environment, and to the invidual experiences of past generations. Wars with the Moors, the Spaniards and the French; the hardships and privations of the early voyages of discovery; rural dramas and tragedies, exploits and adventures: such are the themes which we might expect in Portugal. Yet nothing could be further from the truth. In their place we find the same stuff as everywhere else in Europe, the same fantastical fairy-tales, bearing no relation whatsoever to the life of the people, their setting indeterminate in time and place. The collections of Theophilo Braga, Consiglieri Pedroso, Bernardino Barboso and others are full of easily recognisable variants of *Cinderella, Beauty and the Beast, The Swan Maiden* and those equally well-known tales to which may be given the generic names of *The Supplanted Bride* and *The Calumniated Wife.*

Perhaps, on reflection, this is less surprising than might appear at first sight. Those who are hard put to it to wrest a livelihood from the soil may well seek relief from their cares in the rich pageantry of the fairy-tale. Just as the city typist forgets the drab monotony of her own life in the "high life" of the feuilleton, so do the Portuguese peasantry find an escape from the struggle for existence in the glittering *historias da Carochinha* (tales of Cockaigne).

Devoid of sufficient creative imagination to invent their own plots, they take these where they can find them, that is to say in the common stockpot of Indo-European myth, legend and fairy-tale. After all, these stories meet the most exacting requirements in the matter of melodrama. The hero, if not always well born, is invariably highly deserving, and the villain as wicked as could be wished. Virtue is rewarded even beyond its deserts (if that were possible), and vice is savagely punished.

So it comes about that in Portugal we renew our acquaintance with all those familiar motives which enchanted our childhood: the Fortunate Youngest Son, Magic Gifts and Impossible Tasks (in assorted trios), the Outcast Child, the Friendly Beast and so forth. But with centuries of use, these stories have worn a little thin.

> Quem conta um conto
> Acrescenta um ponto

("Who tells a tale adds on a little") runs the proverb. But most of these stories have lost rather than gained in the telling. Their plots have become confused and their original meaning lost. Several different tales are frequently entangled in an inconsequent and illogical whole.

Here is a characteristic example, more coherent than many, translated from the exact words in which it was dictated to Leite de Vasconcelos by an old peasant woman:

There was once a king who used to go about at night in order to listen at the doors to what people said of him. He passed by an office (*sic*) and heard women's voices. He listened and this is what he heard.

"If only I could marry the king's cook and eat all the tit-bits," said one of these voices.

"For my part", said another, "I would rather marry his butler."

"And I", said the third, "would rather marry the king, for I should bear him three children, two boys and a girl, each with his little star of gold upon his head."

The king ordered the number of the room to be noted and went away. The next day he sent for the three girls and asked them:

"Which of you said yesterday that she would like to marry my cook?" And to this one he said: "Well, you shall marry my cook."

And he also fulfilled the wishes of the other two, which he had heard them express.

Since the first two judged that the third (who was the youngest) must be the happiest, they began to feel very angry with their sister.

The king's wife found herself with child. And when her time came she brought two boys into the world, each with a little golden star upon his head.

The sisters, seizing this opportunity, substituted two dogs for the boys and put the two children into an osier-basket and threw them into the river.

There was a nobleman who was very fond of boating on the river where the osier-basket floated ashore. Seeing this cradle from the boat he ordered it to be taken out of the water and saw the two lovely children. He was delighted with them and took them to his palace. He was a bachelor.

When the king heard that his wife had been delivered, he asked what she had brought forth. It was with horror that he heard the answer that two dogs had been born to the queen. He was very sad, but he bore it and put it out of his mind because of the great friendship which he had for the queen.

Some time afterwards the queen bore a beautiful little girl with a little golden star upon her head. The little girl suffered the same fate as her brothers.

The three grew up and were educated in the nobleman's palace. At length the latter died and left everything to the three children. One day a black woman came begging for alms, but they gave her none. So the black woman said:

"Then you shall not know where are the parrot which says everything, the tree which sings and the golden fountain."

So they gave her alms, and the black woman explained everything to them, saying that they would find a black ointment in a porringer and would see many horses and mares disporting themselves in the woods; that they must approach and cut off a branch of the tree, catch a drop from the golden fountain and bring back the parrot; and that when they came down even though they might hear great noises or harmonies or scolding or voices of any sort, they must not look back. With the ointment they must anoint the horses and mares.

The next day one of the brothers set off. He came to the place and did all that the black woman had told him. But on the way back, hearing cries and shouts, music and songs, he looked back. He was instantly changed into a horse.

The next day the brother and sister awaited him and seeing that he did not return, one of them made himself ready and left. He met the same fate.

The girl also set off. After having cut the branch from the tree, caught a drop of the golden fountain in a bottle, fetched the parrot and anointed all the horses and mares, she was on her way back when all these noises were again heard. But she resisted and did not turn her head. Then all the horses were changed into men and all the mares into women. Then she lived with her brothers and they were happy. The rumour went round that in the children's palaces were a parrot which said everything, a tree which sang and a golden fountain.

When they told the king that the queen had brought forth a bitch, he never wished to see her again, and ordered a pit to be dug in which the queen should be buried up to her waist, and that whoever passed should spit in her face. She, poor thing, was suffering this fate, when the king, out hunting, and hearing of the extraordinary things which were in the children's palace, went to visit them.

At dinner the parrot was at table. Every time the king spoke, the parrot burst out laughing.

The king invited the children to dine at the palace another day. They accepted but asked if they might bring the parrot. The day came and the children went. They passed the pit where the queen was, and the king did all that he could to make the children spit in the poor woman's face. They refused. At dinner the parrot was also at table, and laughed whenever the king spoke. The king asked why. Then the parrot said:

"I am laughing because the king is talking to his children, but does not know them."

Afterwards the parrot explained everything, saying that they should look and see if a star was to be found on their heads. And indeed there was one. Then the king felt a certain remorse, embraced his children and sent for the innocent queen whose pardon he begged.

The sisters were burnt alive.[1]

What is the origin of these tales which to-day constitute the stock-in-trade of the peasant story-teller not only from one end of Europe to the other, but also in other con-tinents? Many theories have been put forward in answer to this question. The fairy-tale has been derived in turn from Ancient Greek and Indian myths, from the religious conceptions of those "Aryan" tribes of whom so much is heard and so little known, from the movements of the sun, the moon and the stars, from the social conditions of an almost unbelievably primitive stage of society when kingdoms were inherited through marriageable daughters (matriarchy) and youngest sons, from the ritual of cults almost equally remote in time, and lastly from the unconscious promptings of suppressed Freudian complexes. Each school of thought is more exclusive, more contemptuous of its pre-

[1] J. Leite de Vasconcelos, *Ensaios Ethnographicos*, II, p. 93.

decessors, than the last. The probability is that each has stumbled upon a fraction of the truth, but none by itself furnishes an adequate explanation of the fairy-tale.

From Haggerty Krappe's *Science of Folklore* a composite definition of the fairy-tale may be taken which reflects the modern tendency to seek its origin primarily in the universal fondness for a story and the almost universal inability to invent a new one. The fairy-tale, then, may be regarded as "a definite type of popular fiction primarily designed to please and entertain...like modern fiction drawing its materials from various sources, usually in the form of motives combined so as to form an organic whole". It "has a definite origin in a definite country and in a definite time to be ascertained for each combination by the historic-geographic method".[1] These motives, the number of which has been estimated at over ten thousand, can be shown to be drawn in most cases from the various sources, most of them of great age, to which reference has been made above. It is in the highest degree unlikely that any of those which figure in the Portuguese tales are of local origin. Nor is it very probable that the historic-geographic method would justify a claim that Portuguese story-tellers have evolved from them any outstandingly original permutations and combinations.

Indeed, one may look in vain for the faintest hint of local colour in the details of setting and characterisation. On the other hand these stories offer the same support, neither more nor less, as the folk-literature of any other country, for the theories of the "anthropological" school of thought who see in the fairy-tale the scarcely distorted reflection of a prehistoric age. If, in the story of *A Vaquinha de Ouro* collected by Bernardino Barboso near Évora, a king wants to marry his daughter, we are free either to assume or to doubt that this motive reflects a royal and ritual incest. And if, in *As Tres Cidras de Amor*, a chair is made from the bones of the villainess and a drum from her skin, it must be left to common-sense to decide whether in some bygone age these unusual materials occasionally took the place of wood and parchment. On the other hand the savagery with which vice is habitually punished

[1] Alexander Haggerty Krappe, *op. cit.* p. 144.

in these tales has a very primitive character and accords ill with the humane outlook of the present-day Portuguese.

These fairy-tales reveal markedly little sense of probability, of proportion or of humour. This is reserved for another branch of the folk-tale, one which is usually briefer and may be called the ancestor of the modern humorous anecdote, to wit the "merry tale".

A poor devil who had departed this world, runs one such tale, presented himself one day at the gates of Paradise.

"What do you want here?" called St Peter from his watch-tower.
"To enter Heaven."
"Have you passed through Purgatory?"
"Not exactly, but I was a married man on earth."
"That comes to the same thing. You may enter."
A little later another soul appeared, and the following conversation took place:
"What do you want here?" asked St Peter.
"To enter Heaven."
"Have you passed through Purgatory?"
"No, but you have just let a man in, who came straight from earth."
"Yes, but he was married."
"Well then, I must have a double claim to enter, for I was married twice."
"Get along with you, you rascal. Do you think Heaven was made for fools?"

Another type of story well represented in Portugal is the animal tale. Most of these are derived from Aesop's Fables and kindred sources, but the bailiff of an Alentejan *monte* tells one with which few will already be familiar:

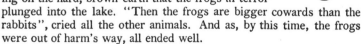

One day, all the animals gathered together beside a lake to decide which of them was the timidest, for the world was no place for the faint-hearted, they declared, and these must be put to death. A vote was taken, and it was decided that the rabbits were the most timorous of all creatures. When they heard this all the rabbits began to tremble for fear. Such a noise did they make with their little quivering paws pattering on the hard, brown earth that the frogs in terror plunged into the lake. "Then the frogs are bigger cowards than the rabbits", cried all the other animals. And as, by this time, the frogs were out of harm's way, all ended well.

A shrewd wit may be found in a series of tales centring round the apocryphal wanderings of Our Lord and St Peter. Some of these are identical with those told in the Basque Country. Others were new to me, notably the following:

> Our Lord and St Peter were walking along a road when they passed a ploughman who was swearing horribly. Jesus said to him: "Good day to you, son of the Lord." Further on they passed a poor man squatting on the ground telling his beads. To him Jesus said: "Good day to you, son of the Devil." St Peter, wondering, asked the reason for these words. Jesus Christ replied that the ploughman was working with all his heart to support his family, while the poor man was planning how he might go and rob him.[1]

> On another occasion, when Our Lord and St Peter were going begging through the world, they found four little pigs on their road and gave them to a woman to look after. Our Lord said to the woman: "Take care of these four little pigs, and when we come back in a year's time we will share them, two for you and two for me." The pigs grew so that they were a wonder to all. The year passed, and Our Lord and St Peter returned. But the woman had hidden two of them, intending to account only for the other two. When Our Lord asked for them, she replied: "Two died, but here are the remaining two." Then Our Lord answered: "These two which are here shall be yours and mine, but the two which are shut away in the sty shall run wild over the mountains." And from these are descended the whole race of wild boars.[2]

This story is of the type known as "aetiological". That is to say, it was invented to account for something of which the true cause or origin was unknown. There are many such tales in Portugal explaining matters so far apart as the creation of the world, the quail's inability to perch, place-names like Braganza and Viseu (good examples of *Volksetymologie*) and the lack of nourishment in the seeds of the *tremoço* lupin. The rattling of the latter in their pods, runs the legend, startled Our Lady on her flight into Egypt, and she laid a curse upon them that they should fill but not sustain.

Last of all comes the "local" legend; local in its immediate application, but often as widely distributed as any other and attributed to half a dozen different places in as many countries.

---

[1] J. Leite de Vasconcelos, *Ensaios Ethnographicos*, II, p. 206.
[2] *Ibid.* p. 93.

One such tale is that of the deaf Ribatejano quoted earlier in this book. Another was told to me at Braganza:

Two young men, a Portuguese and a Spaniard, studied theology together at the University of Salamanca. The Spaniard constantly proclaimed his intention of becoming Archbishop of Salamanca, while the Portuguese was content with the more humble ambition of becoming the Priest of his own village, Quintanilha. Many years later, having attained this ambition, the Portuguese had occasion to visit the tiny hamlet of San Martiño, just across the Spanish border. Here, to his surprise, he recognised in the ragged sacristan none other than his old fellow-student. "What then has become of all your great ambitions?" he asked the latter. "Had it not been for my ambitions", was the dignified reply, "I should never have risen so high."

In conclusion, a word of mention is due to the traditional tags with which Portuguese folk-tales frequently conclude. These usually take the form of a rhyming couplet such as:

Vitória, vitória!	Victory, victory!
Acabou a historia.	My story's done.

Strangely enough, some seem to presuppose an occasional incredulous listener:

Quem o contou 'qui está:	He who told it is here:
Quem o quiser saber va lá.	He who would make sure may go there.
A certidão está em Tondela,	Certainty is in Tondela,
Quem quiser va lá por ela.	Who wills may go and seek her.

Of the proverb there is from the general point of view little new to be said. It is the formula into which are condensed, simplified and generalised all human thought and experience, the wise and the foolish, the learned and the popular, the moral and the cynical. Do the proverbs of a race exhibit its character? In the main they do not, for although "each proverb was coined just once, in a given locality, at a given time by one mind with some gnomic talent",[1] this act of creation took place, in most cases, before the formation of existent nationalities. A large proportion of the proverbs which are so widely disseminated throughout Europe to-day can be traced back to Latin or Semitic sources.

[1] Haggerty Krappe, *op. cit.* p. 144.

Nevertheless, the proverb cannot be omitted from a survey of Portuguese folk-lore, for it plays at least as large a part in the Portuguese as in any other folk-literature.

"Old words are wise words", say the Basques. The Portuguese peasants subscribe to this touching faith in proverbial wisdom by calling them "little Gospels". That antiquity has always been regarded as their hall-mark is confirmed by the expression *vervo antiguo* under which the oldest recorded examples have come down to us in the mediaeval *Cancioneiros*. There is one in the *Cancioneiro da Ajuda*:

E porem diz o vervo antiguo:	Indeed the ancient adage runs:
A boy velho non busques abrigo.	From an old ox no refuge need be sought.

By the sixteenth century the name had generally changed to *exempro antigo* (recalling the modern expression *exemplo da velha*, old wives' saw) in which form it figures in many of Gil Vicente's *autos*:

Porque diz o exempro antigo:	For the ancient adage has it:
Que a amiga e o amigo	The lover and the beloved
Mais aquenta que bom lenho.	Blaze more hotly than tinderwood.

The first anthology of Portuguese proverbs was compiled by one Antonio Delicado and published in Lisbon in 1651 under the title of *Adagios Portuguêses reduzidos a logares Communs*. This collection of proverbs, many of which are still in current use, was the first of a number formed from both literary and oral sources.

In these old saws it is less the substance than the form which is of interest to-day. Much of their philosophy is trite, hackneyed, sententious, superficial and even contradictory. Thus unreliability, blamed in one proverb

Palavras sem obras,	Words without deeds,
Citara sem cordas,	A zither without strings,

is excused in another

Dizer e fazer não é para todo o homem,	Saying and fulfilling is not for every man,

and approved in a third:

Quem não mente	Who tells no lies
Não vem de boa gente.	Comes not of decent folk.

Many of our old, familiar home-truths are restated in a new form, with the aid of metaphors which are striking in themselves and bring out more than does their meaning the local flavour of Portugal. "Too many cooks spoil the broth" has several equivalents:

A cera sobeja Queima a igreja.	Too many candles Burn down the church.
Muitas enfeitadoras estragam a noiva.	Too many adorners spoil the bride.
Nem tantos ámens que se dana a Missa.	Not so many amens or you'll spoil the Mass.

"Once bitten, twice shy" is rendered by:

Gato a quem morde a cobra tem medo á corda;	A cat that a snake has bitten is afraid of a rope;

or, more indirectly, by

Ver as barbas do vizinho arder e pôr as suas de molho.	See your neighbour's beard catch fire and moisten your own.

A great number of these proverbs are concerned with religion, though by no means all such are moral sentences. If some breathe a spirit of trusting faith:

Mais vale quem Deus ajuda Do que quem muito madruga,	God's help is of more avail Than early rising,

others are frankly sceptical:

Não há que fiar em Deus em tempo de inverno.	In winter time count not on trust in God.

Among them are many which, without reflecting an agnostic outlook, ridicule too implicit or too material a faith in saintly invocations:

Quando Deus não quer santos não rogam.	Against God's will no Saints will mediate.
Quando há que comer em casa, sãos estão os santos.	When there is food in the house the Saints are left in peace.
O rio passado, O santo não lembrado.	The river crossed, The Saint forgotten.
Para baixo todos os santos ajudam.	Downhill, all Saints help.

Priests and friars are as usual favourite objects of popular satire:

Sol madrugador, homem rezador e frade cortês, Arrenega-me de todos três.	Morning sun, praying men and courteous monks, I forswear all three.
Quem pede para a candeia Não se deita sem ceia.	Who begs oil for the altar Goes not hungry to bed.
O ladrão que anda com o frade Ou o frade será ladrão Ou o ladrão será frade.	The thief consorts with the monk, The monk will be a thief Or the thief become a monk.
Clerigo que foi frade Nem por amigo nem por compadre.	A priest turned monk Neither for friend nor companion.

No more than priests are women spared. A number of proverbs compare them, on differing grounds, with hens:

A mulher e a galinha, com sol a casa.	Women and hens should be home by sundown.
Á mulher e á galinha, torcer-lhe o colo se a queres fazer boa.	Women and hens, 'tis wringing their necks that makes them good.

Many of these proverbs afford glimpses into the past, recalling the great discoveries of the sixteenth century:

Á India mais vão do que tornam;	To India more go than return;

and the Inquisition in the seventeenth:

Com Elrei e com a Inquisição Chitão.	With the King and the Inquisition Mouth shut!

The searing East wind, scorching in summer and frozen in winter, is linked in the popular mind with the misfortunes which dynastic marriages brought upon Portugal:

De Espanha nem bom vento Nem bom casamento,	From Spain, nor good wind Nor good marriage,

and the Peninsular War has also passed into proverb:

Quartel Geral em Abrantes, Tudo como dantes.	Headquarters at Abrantes, Everything as before.

Other sayings are of folkloric interest. The rowdy side of the *romarias* is well brought out:

Ás romarias e ás bodas Vão as sandias todas.	To *romarias* and weddings Go all giddy maidens.
Boa romaria faz Quem em sua casa está em paz.	He makes a good pilgrimage Who stays in peace at home.
Depois da pança Vem a dança.	When the belly's full Then comes the dance.

Two contain allusions to the slaughter of the pig on St Martin's or St Thomas' Day, which is still performed in very much the same manner as shown in the *Book of Hours* of King Manoel (1517):

Cada porco tem seu São Martinho.	Every pig has his Martinmass.
O dia de São Tomé, quem porco não tiver, Matar pode a mulher.	On St Thomas' Day Who has no pig his wife may slay.

The superstitious prejudice against bearded women and red-headed men is preserved in rhyme:

Homem ruivo e mulher barbuda De longe se sauda.	Red-headed man and bearded woman, Greet them from afar.
Á mulher barbuda Não dês pousada.	To a bearded woman Give no shelter.

In the *terras de Miranda* the privilege of shouldering in the annual procession the stand on which the image of a Saint is set is often put up to auction for the benefit of the Church. Hence the proverb:

Quem mais der, mais amigo é do Santo.	Who gives most is the Saint's best friend.

It is difficult to know whether it is the medicinal or the magic properties of garlic which have given rise to the saying:

Quem come alhos com casca Dá pancada que lasca.	Who eats garlic with its skin Can strike a blow which makes its mark.

Like the quatrains, many of the proverbs are of local application. The "three infernal months" of summer are recalled by one which I heard at Miranda do Douro:

Aguas de verão	Summer rains
As moscas as bebam.	The flies may drink them.

Another, of which the allusion is obscure, reflects unfavourably on the southernmost province of Portugal:

Mais passou Nossa Senhora no Algarve.	Worse fared Our Lady in the Algarve.

And worse than Our Lady in the Algarve fare the Spanish Galicians in Portuguese proverbs. They would appear to be a byword for everything despicable:

Duzentos galegos não fazem um homem	Two hundred Galicians do not make a man
Senão quando comem.	Save when they eat.
Guarda-te de cão preso	Beware of tethered dog
E de moço galego.	And Galician lad.

But malice, like charity, begins at home, and many Portuguese towns and villages with their inhabitants are also singled out for insult or ridicule:

Aviz	Aviz
Que nem o Diabo quis.	Which not even the Devil wanted.
Cascais	Cascais
Uma vez e nunca mais.	Once but never again.
Meninas de Montemor, com Deus me deito!	Maidens of Montemor, I'll lie with God!
Quem burro vae a Santarem	The ass who goes to Santarem,
Burro vae e burro vem.	Ass he goes and ass returns.
Guarde-o Deus lá no Barreiro.	God protect you, over there in Barreiro.
Negrões	Negrões
Doze moradores e treze ladrões.	Twelve inhabitants and thirteen thieves.

"The thirteenth is the village priest", the peasants are apt to add.

The accumulated experience of countless generations is preserved in proverbial lore for the benefit of their descendants. Sound medical and gastronomical counsel is administered in tabloid form:

Depois de melão Vinho de tostão; Depois da melancia Água fria.	After melon A pennyworth of wine; After water-melon Cold water.
Não comas quente, Não perderás o dente.	Eat no hot food, You'll lose no teeth.
Almoçar com um caçador, jantar com um lavrador e cear com um arrieiro.	Breakfast with a sportsman, lunch with a farmer and sup with a muleteer.
Leitão de mês E cabrito de três.	Sucking-pig at one month And roast kid at three.

"And maidens from eighteen to twenty-three," add the gayer sparks.

There is also the usual seasonal and weather lore, condensed into mnemonic jingles:

Lua nova e lua cheia Praia mar ás duas e meia.	New moon and full moon High tide at half-past two.
Águas verdadeiras S. Mateus são as primeiras.	Of real rains The first are on St Matthew's Day.
Passado o Natal O dia cresce um passo de pardal.	Christmas past The days lengthen by a sparrow's hop.
Ruivas ao Nascente Desapõe os bois e foge sempre; Quando estão as ruivas ao mar Pega nos bois e vai lavrar.	Red in the East Always unyoke your oxen and flee; When the red is over the sea Take your oxen and go and plough.

Agriculture is also to the fore:

Por S. Martinho Semeia fava e linho.	On St Martin's Day Sow beans and flax.
Fevereiro quente Traz o diabo no ventre.	A warm February Carries the Devil in its belly.
O milho pelo São João Deve cobrir um cão.	By St John's Day The maize should cover a dog.

Vinhas,	Vines,
Minhas;	My own
Olivais,	Olives,
Dos meus pais;	My parents';
Montados,	Cork woods,
Dos meus antepassados.	My forbears'.

This last is an allusion to the varying time taken by the different cultivations to yield.

For every little incident the Portuguese seem to have an appropriate proverb. A Lisbon friend one day heard terrible cries coming from a house which he entered to find a man attacking his wife with a table-fork. He defended the woman and was in due course called upon to give evidence at the ensuing trial for assault. To his surprise the wife met him at the entrance to the court and begged him not to bear witness against her husband. "Don't you know the proverb?" she asked: *Entre marido e mulher, não se meta nem uma colher* ("Between husband and wife set not so much as a spoon"). "Yes," replied my friend, laughing, "but what about a fork?"

No better conclusion could be found to this chapter, and indeed to the whole book, than a further brief selection from the store of proverbs which, whatever their origin, have come to sum up the lives of the Portuguese people, their joys, their sorrows, their unending endeavour, transitory in each separate individual, but rendered universal and eternal by their endless recurrence in all mankind:

Cada roca com seu fuso, Cada terra com seu uso.	To every distaff its spindle, To every land its ways.
O homem fogo, a mulher estopa, Vem o diabo e assopra.	Man is fire and woman tow, The Devil comes and starts to blow.
A panela de vinho é o melhor juiz de paz.	The bowl of wine is the best Justice of the Peace.
Quem cospe para o céu na cara lhe cai.	Who spits at Heaven it will fall back in his face.
O que não mata engorda.	What does hot kill fattens.
Não sirvas a quem serviu Nem peças a quem pediu.	Serve not him who once has served; Beg not from him who once has begged.

BIBLIOGRAPHY
OF WORKS QUOTED OR CONSULTED

ALMEIDA GARRETT, J. B. da S. L. Romanceiro. 3 vols. Lisbon, 1851–3.
ALVES, PADRE FRANCISCO MANUEL. Trás-os-Montes. Lisbon, 1930.
—— Chaves. Gaia, 1931.
A.P.D.G. Sketches of Portuguese Life, Manners, Costume and Character. London, 1826.
ARROIO, ANTONIO. A Canta Coral e a sua Funcção Social.
ATHAIDE OLIVEIRA, FRANCISCO XAVIER DE. As Mouras Encantadas e os Encantamentos no Algarve. Tavira, 1898.
—— Contos e Tradições do Algarve. 2 vols. Tavira, 1900–5.
—— Romanceiro e Cancioneiro do Algarve. Oporto, 1905.
AUBRY, PIERRE. Trouvères et Troubadours. Paris, 1909.
AURORA, CONDE DE. Roteiro da Ribeira Lima. Ponte de Lima, 1929.
BARROS, GAMA. Historia da Administração Pública em Portugal nos Seculos xii a xv. Lisbon, 1885–1922.
BASTO, CLAUDIO. Traje á Vianesa. Gaia, 1930.
BATALHA, LADISLAU. Historia Geral dos Adágios Portugueses. Lisbon, 1924.
BATALHA REIS, PEDRO. Da Origem da Música Trovadoresca em Portugal. Lisbon, 1931.
BECKFORD, WILLIAM. Travel Diaries. 2 vols. London, 1928.
BELL, AUBREY F. G. In Portugal. London, 1912.
—— Poems from the Portuguese. Oxford, 1913.
—— Studies in Portuguese Literature. Oxford, 1914.
—— Lyrics of Gil Vicente. Oxford, 1914.
—— Portugal of the Portuguese. London, 1915.
—— Portuguese Portraits. Oxford, 1917.
—— Gil Vicente. Oxford, 1921.
—— Portuguese Literature. Oxford, 1922.
—— Portuguese Bibliography. Oxford, 1922.
—— (Edited by). The Oxford Book of Portuguese Verse. Oxford, 1925.
BRAGA, THEOPHILO. Romanceiro Geral. Coimbra, 1867.
—— Cancioneiro Geral. Coimbra, 1867.
—— Historia da Poesia Popular Portuguez. Oporto, 1867.
—— Floresta de Varios Romances. Oporto, 1868.
—— Contos Populares do Archipelago Açoriano. Oporto, 1869.
—— O Povo Portuguez nos seus Costumes, Crenças e Tradições. 2 vols.
—— Contos Tradicionaes do Povo Portuguez. 2 vols. Oporto.
—— Romanceiro Geral Portuguez. 3 vols. Lisbon, 1906–9.

Cancioneiro da Ajuda. (Edited by Carolina Michaelis de Vasconcelos.) 2 vols. Halle, 1904.

Cancioneiro Colocci-Brancuti. Halle, 1880.

Cancioneiro Geral. Lisbon, 1516.

Cancioneiro da Vaticana. Halle, 1875.

CASTRO, JOÃO BAUTISTA DE. Mappa de Portugal. Lisbon, 1749.

CHAVES, LUIS. Chaminés de Portugal. Famalicão, 1929.

—— Os Pelourinhos Portugueses. Gaia, 1930.

—— Trás-os-Montes. Gaia, 1931.

—— Portugal Além. Gaia, 1932.

COELHO, FRANCISCO ADOLPHO. Contos Populares Portuguezes. Lisbon, 1879.

—— Os Ciganos de Portugal. Lisbon, 1892.

CONSIGLIERI PEDROSO, ZOPHIMO. Tradições Populares Portuguezes. Lisbon, 1882.

—— Portuguese Folk Tales. Translated by Miss Henriqueta Monteiro. London, 1882.

CORREIA LOPES, EDMUNDO ARMENIO. Cancioneirinho de Fozcoa. Coimbra, 1926.

CORTESÃO, JAIME. Cancioneiro Popular. Oporto, 1914.

DELICADO, ANTONIO. Adagios Portuguêses reduzidos a logares Communs. Lisbon, 1651.

DEUSDADO, DOMINGOS FERREIRA. Trás-os-Montes. Lisbon, 1930.

DEUSDADO, FERREIRA. Escorços Trasmontanos. Lisbon, 1912.

FELGUEIRAS, GUILHERME. Espadeladas e Esfolhadas. Gaia, 1932.

FERNANDES TOMÁS, PEDRO. Canções Populares da Beira. Coimbra, 1923.

—— Velhas Canções e Romances Portuguêses. Coimbra, 1923.

FERREIRA, JOSÉ MARIA DE ANDRADE. Curso de Literatura Portuguêsa. Lisbon, 1875.

FIGUEIREDO, ANTERO DE. Jornadas em Portugal. Lisbon, 1921.

Folklore y Costumbres de España. Edited by F.? Carreras y Candi. 2 vols. Barcelona, 1931.

FRAZER, SIR JAMES. The Golden Bough. Abridged edition. London.

FREITAS BRANCO, LUIS DE. A Música em Portugal. Lisbon, 1930.

GOMES, JOSÉ. O São João em Braga. Braga, 1904.

GRAÇA, A. SANTOS. O Póveiro. Póvoa de Varzim, 1932.

GUIMARÃES, FELICIANO. Azulejos de Figura Avulsa. Gaia, 1932.

GUIMARÃES, LUIS D'OLIVEIRA. Os Santos Populares. Gaia, 1931.

HAGGERTY KRAPPE, ALEXANDER. The Science of Folklore. London, 1930.

HARDUNG, VICTOR EUGENIO. Romanceiro Portuguez. 2 vols. Leipzig, 1877.

JACKSON, CHARLOTTE LADY. Fair Lusitania. London, 1874.

JEAN-JAVAL, LILY. Sous le Charme du Portugal. Paris, 1931.

Leça, Armando. Da Música Portuguêsa. Lisbon, 1922.

Leite de Vasconcelos, José. As Maias. Lisbon, 1882.

—— Tradições Populares de Portugal. Oporto, 1882.

—— Poesia Amorosa do Povo Portuguez. Lisbon, 1896.

—— Estudos Ethnographicos. 4 vols. Lisbon, 1896–1920.

—— Religiões da Lusitania. 3 vols. Lisbon, 1897–1912.

—— Estudos de Philologia Mirandesa. 2 vols. Lisbon, 1900.

Lopes Dias, Jaime. Etnografia da Beira. 3 vols. Lisbon, 1926–9.

Malheiro Dias, Carlos. Cartas de Lisboa. Lisbon, 1905.

Martins, Padre Firmino A. Folklore do Concelho de Vinhais. Coimbra, 1928.

Michaelis de Vasconcelos, Carolina. Tausend Portugiesische Sprichwörter. Brunswick, 1905.

—— Estudos sobre o Romanceiro Peninsular. Madrid, 1907.

Murray, Margaret Alice. The Witch Cult in Western Europe. Oxford, 1921.

Neves, Cesar das and Campos, Gualdino de. Cancioneiro de Músicas Populares. Oporto, 1893.

Neves e Mello, Adelino Antonio das. Músicas e Canções Populares Colligidas da Tradição. Lisbon, 1872.

Pereira, José Manuel Martins. As Terras de Entre Sabor e Douro. Setúbal, 1908.

Picão, José da Silva. Atraves dos Campos. Elvas, 1903.

Pimentel, Alberto. A Triste Canção do Sul. Lisbon, 1904.

—— As Alegres Canções do Norte. Lisbon, 1905.

Pinho Leal, Augusto. Portugal Antigo e Moderno.

Pinto, Alfredo. Em Terras de Portugal. Lisbon, 1914.

Pinto de Carvalho (Tinop). Historia do Fado. Lisbon, 1903.

Pires, Antonio Thomaz. Cancioneiro Popular Politico. Elvas, 1890.

—— Cantos Populares Portuguezes. 4 vols. Elvas, 1902–10.

Rey Colaço, Alexandre. Cantigas de Portugal. Lisbon, 1922.

Ribeiro Fortes, José Maciel. O Fado. Oporto, 1926.

Ribera y Tarragó, Julián de. La Música de las Cantigas. Madrid, 1922.

Rocha, Abilio Monteiro e Sousa. Poesias e Canções Populares do Concelho de Maia. Oporto, 1900.

Rodrigues Lapa. Das Origens da Poesia Lírica em Portugal na Idade Media. Lisbon, 1929.

Sampaio, Albino Forjaz de. Historia da Literatura Portuguesa. Lisbon.

Saraiva e Castilho. A Ribeira Saraiva. London, 1877.

Seara, Francisco José Ribeiro. Bosquejo Historico da Villa de Vallongo e Suas Tradições. Vallongo, 1896.

Sitwell, Sacheverell. Southern Baroque Art. London, 1924.

—— Spanish Baroque Art. London, 1931.

SMITHES, M. F. Things Seen in Portugal. London, 1931.

STEPHENS, H. MORSE. Portugal, 1908.

TEIXEIRA, ANTONIO JOSÉ. Em Volta duma Espada. Oporto, 1930.

UNAMUNO, MIGUEL DE. Por Tierras de Portugal y de España. (Volume 9 of Collected Works.) Barcelona, 1930.

URTEL, HERMANN. Beitrage zur Portugiesischen Volkskunde. Hamburg, 1928.

VEIGA, SEBASTIÃO PHILIPPE MARTINS ESTACIO DA. Romanceiro do Algarve.

VIANA, ABEL. Estações Paleolíticas do Alto Minho. Oporto, 1930.

—— Notas Historicas, Arqueológicas e Etnográficas do Alto Minho. Viana do Castelo, 1930.

VICENTE, GIL. Obras. Hamburg, 1834.

VINDEL, PEDRO. Martin Codax. Las Siete Canciones de Amor. Madrid, 1915.

VITERBO, SOUSA. Artes e Artistas em Portugal. Lisbon, 1920.

VITORINO, PEDRO. Cerâmica Portuense. Gaia, 1930.

WATSON, W. CRUM. Portuguese Architecture. London, 1908.

YEARSLEY, MACLEOD. The Folklore of Fairy Tale. London, 1924.

YOUNG, GEORGE. Portugal, Old and Young. Oxford, 1917.

—— (Edited by). Portugal: An Anthology. Oxford, 1916.

PERIODICALS

A Arte Musical. Lisbon.

Almanach de Lembranças. Lisbon.

Alma Nova. Lisbon.

Archeologo Português. Lisbon.

Folk-Lore. London.

Lusa. Viana do Castelo.

Lusitania. Lisbon.

Modern Language Review. Cambridge.

Portucale. Oporto.

Portugalía. Oporto.

Revista de Ethnologia e de Glottologia. Lisbon.

Revista de Guimarães. Famalicão.

Revista Lusitana. Lisbon.

Revista do Minho. Espozende.

Revue Hispanique. New York, Paris.

Trabalhos da Sociedade Portuguesa de Antropologia e Etnologia. Oporto.

INDEX

Abengayats, 198
Abenhabib, 198
Abiul, 25
Abrantes, 276
Abrantes, Ventura de, 247, 250
Adufe, 201, 214
Afife, 41
Afonso III of Portugal, 14, 15
Agueda, River, 3
Aguiar da Beira, 31
Alcabideche, 104
Alcobaça, 22, 23, 24, 79
Alde Galega, 118
Aldeia Nova de Ficalho, 231
Alenquer, 104
Alentejo, 9–13, 62, 71, 78, 92, 120,
    123, 124, 148, 179, 202, 206, 208,
    231, 239, 271
Alfama, 7, 140–2, 245, 252
Alfonso VII of Spain, 192, 195
Alfonso the Sage, 196
Alford, Violet, 168
Algarve, 13–16, 58, 59, 62, 70, 71,
    86, 87, 95, 123, 124, 127, 162,
    198, 206, 249, 278
Aljubarrota, 23
Aljustrel, 128
Almada, 7, 47
Almeida, 31
Almeida Garrett, 215
Almurtão, 131, 156, 208
Alter do Chão, 117
Alvações do Corgo, 124, 125
Alvares, Nuno, 23, 217
Amarante, 131, 133, 166
Amazons, 81
Amendoeira, 138
Amulets, 60–6, 124–5, 146, 151, 238
Andalusia, 12, 14, 197, 249
Andersen, Hans, 21
Angola, 106
Animals, lore relating to, 55, 56, 60,
    64, 65, 81–3, 103–4, 116–18, 125,
    128–9, 138–9, 175, 271, 272
A.P.D.G., 106, 132, 134, 137, 206
Apúlia, 23
Arbeau, 175
Arce, F. de, 175
Architecture, 12, 17–18, 21–2, 23,
    28, 30, 42, 243
Arcos de Valdevez, 80
Arcozelo da Serra, 85, 166, 175

Ardennes, 120
Areosa, 41
Arganil, 116, 146
Arrábida, Serra da, and Monastery of,
    8, 128, 131, 136
Arraiolos, 10
Arroio, A., 248
Ascension Day, 148
Astorga, 39
Asturias, 58
Atalaia, 128, 131, 161–2
Aubry, Pierre, 195
Aurora, Condes de, 102
Auto de Floripes, 176–9
Ave, River, 44, 77, 156
Aveiro, 23, 29, 135, 163, 251
Avelar, 25
Avila, 23
Aviz, 278
Ayamonte, 14
Azeitão, 8
Azores, 54, 59, 76, 99, 134, 176
Azulejos, 42–3

Baçal, Abade, 160
Bagpipes, 170, 201, 238
Ballads, 193, 207–8, 215–29
Barbieri, 196
Barboso, Bernardino, 266, 270
Barcelos, 42, 85, 157, 176
Barcelos, Conde de, 190
Barlavento, 15
Barreiro, 7, 278
Barroso, Terras de, 39, 43, 77, 86,
    92, 94, 96, 120, 122, 147, 185
Basque folklore, parallels with, 56,
    79, 81, 96, 102, 103, 115, 120, 129,
    170, 175, 176, 216, 224, 248, 272,
    274
Basse-Navarre, 120
Batalha, 22, 23
Batalha Reis, Pedro, 195, 197
Bayonne, 252
Beauharnais, Joséphine, 252
Beckford, William, 6, 17, 118, 253
Beira, 13, 28–9, 120, 131, 135
Beira Alta, 4, 11, 28–31, 52, 92, 99,
    113, 120, 150, 179
Beira Baixa, 28, 31–3, 137, 197, 201,
    203, 208
Beira Mar, 29
Beja, 10, 124, 198

Belas, 53, 127, 182
Belém, 16, 22, 43
Bell, Aubrey, 191, 217, 228, 262
Belloc, Hilaire, 37
Belmonte, 32
Birth, lore relating to, 84–6
Bisalhães, 11
Blackened faces, 107, 113, 183
Boats, 24, 28
Bombarral, 259
Bornes, Serra de, 36
Bouro, Terras de, 92
Braga, 39, 42, 73, 75, 88, 107, 117, 126, 131, 159, 164–6, 171, 175
Braga, Luis de Almeida, 36, 200
Braga, Theophilo, 176, 249, 266
Braganza, 36–7, 38, 93, 98, 103–4, 123, 135, 137, 152, 169, 175, 180, 272, 273
Brancuti, Count Paolo, 189
Breton lais, 217, 218
Broom, 60, 65, 124–5, 153
Brydone, 134
Buarcos, 239
Bull-fighting, 18–21
Bull-roarer, 54
Busaco, 4, 23, 29, 226
Byron, Lord, 231

Cacilhas, 7
Cadaval, 21, 118
Cadriceira, 109
Caldas da Rainha, 11, 23
Caminha, 42
Camões, 22
Campos, Agostinho de, 231
Canas de Senhorim, 29, 180
Cancioneiros, 189–98, 242, 274
Cancionero del Palacio, 196
Caneças, 73
Caparica, 24, 60
Capinha, 157
Caramulo, Serra de, 29, 77
Carnival, 104–18, 123, 146, 164, 168–9
Carrazedo de Bouro, 98
Carreço, 41, 206
Carregal do Sal, 185
Carregôsa, 130
Carregueira, 161
Carregueiras, 63
Carvalho, Pinto de, 245, 247, 249
Casalinhas, 169
Cascais, 104, 132, 278
Caseigas, 211
Castelo Branco, 31, 134
Castelo Branco, Camilo, 35, 241

Castile, 3, 5, 33, 37, 156, 192, 195, 197, 215–16, 228, 239
Castro Daire, 30
Catherine of Braganza, 250
Cats, lore relating to, 65, 82, 116, 146
Cavado, River, 39
Ceia, 33
Cercal, 111
Cercio, 170, 180
Charles II of England, 250
Charms, 65–8
Chaves, 35, 38, 39, 77, 180
Chaves, Luis, 92, 159, 180
Chestnuts, 102, 125, 182
Choupal, 4, 28, 261
Christmas, 100, 180, 182–3, 279
Cintra, 16–17, 29, 58, 89, 127, 136, 137, 145, 182
Circuiting, 137–8
Cirios, 160–2
Citânia de Briteiros, 44, 77, 144, 200
Ciudad Rodrigo, 3, 31
Codax, Martin, 191, 194
Coimbra, 4, 11, 22, 25, 26–8, 29, 107, 131, 161, 164, 182, 224, 243, 247, 254, 261
Colares, 104
Colocci, Angelo, 189
Compostela, see Santiago de Compostela
Constituições, episcopal, 73, 75, 98, 135
Cordova, 14
Corpus Christi, Festival of, 61, 117, 163–4, 175
Correia Lopes, 197, 249
Cortiças, 122
Cossantes, 156, 191–8, 241
Covilhã, 31, 32, 184, 210, 211
Crasto, 179
Cuckoo, lore relating to the, 118
Cumieira, 75
Cyclops, 81

Dances, 109, 115, 141, 145, 149, 163–76, 179, 191–2, 204, 206–8, 241, 248, 251–3, 277
Dantas, Julio, 264
Dão, River, 29
Dead, souls of the, 80, 96–9, 100, 183–5
Death, lore relating to, 86, 94–9, 115, 120
Delicado, Antonio, 274
Devil, the, 57–9, 65, 66, 98, 115, 280
Dezoteux, 252
Diana, 58
Dias, Carlos Malheiro, 253

Dinis, King, 24, 189, 228
Divination, 52, 75–6, 85, 146–7
Dolmens, 77
Douro, River, 30, 31, 33–4, 37, 44, 72, 99, 197, 202, 206, 231, 232
Douro Province, 28
Dragons, 163–4, 174
Duas Igrejas, 180
Durham, Edith, 96

Easter, 100
Eloi, João, 89
Elvas, 10, 60, 111, 131, 132, 208, 231
Emperor, festival of, 104–5
Entradas, 74
Entre-Minho-e-Douro, see Minho Province
Epiphany, see Twelfth Night
Espariz, 107
Espichel, Cape, 8, 160
Espinho, 24
Esposende, 88, 135
Essex, Earl of, 14
Estoril, 16, 108
Estrela, Serra da, 4, 17, 29, 30, 31, 32–3, 43, 180, 213, 232
Estremadura, 7–9, 16–22, 93, 123, 130, 160, 193, 206, 208
Estremoz, 10, 11, 92, 113, 115
Evil eye, 60, 61, 63, 124
Évora, 11–12, 74, 102, 112, 135, 142, 270

Fado, the, 27, 112, 199, 215, 245–65
Fafe, 128
Fairies, 79
Falcão, Cristovão, 243
Falmeno, 254
Famalicão, 118
Fandango, 203, 206, 252, 254
Faro, 14
Fátima, 24, 130
Feira, 149
Ferreira, J. M. de A., 253
Festa dos Rapazes, 103–4, 169, 175
Fig, the, 62, 64
Figueira de Castelo Rodrigo, 31
Figueira de Foz, 239
Figueiredo, Antero de, 38, 43
Flamenco, 142
Flax, lore relating to, 144, 174, 232
Floripes, see Auto de Floripes
Folk medicine, 52, 61–3, 66, 68, 69, 70, 71–2, 73
Fortes, J. M. Ribeiro, 248
Fozcoa, 135
Foz do Dão, 102

Frazer, Sir James, 101, 108, 111, 113–14, 151
Freitas Branco, Luis de, 250
Fuentes de Oñoro, 4
Fundão, 124
Furadouro do Mar, 24

Galicia, 34, 38, 58, 97, 189–96, 214, 239, 278
Gama, Vasco da, 16
Gama Barros, 157
Garlic, lore relating to, 60, 61, 62, 64, 86, 277
Gascony, 120
Gentiles, 81
Gerez, Serra do, 40, 43, 86
Godim, 58
Gojim, 128, 130
Gouby, 120
Gouveia, 32, 33, 166
Gralheira, Serra da, 29
Grandola, 9
Grão Vasco, 30
Guadiana, River, 14, 23, 33
Guarda, 31, 75, 77
Guilds, professional, 163–4, 175
Guimarães, 42, 44, 76, 77, 86, 149, 162
Guimarães, Luis d'Oliveira, 139
Guitars, 27, 202, 245–6, 250
Gypsies, 75, 92, 142–3

Halloween, 100, 121, 181, 182, 184
Hand of Glory, the, 60
Hannibal, 198
Harbord, Charles, 89
Hellenes, 80
Henry the "Navigator", Prince, 16, 22
Herbs, 52, 60, 64, 86, 145, 148–9, 153, 222

Idanha, 208
Inez de Castro, 25, 28, 227
Isabel, Queen, 28, 104, 131

Jackson, Charlotte Lady, 249
Janas, 58, 138
Janeiras, 102–3, 184
Janus, 58
Jarmelo, 92, 93
João II of Portugal, 95
João V of Portugal, 17, 253
José I of Portugal, 253
Judas, burning of, 115–16

King David, dance of, 164–6
Krappe, Alexander Haggerty, 270

Lacerda, Francesco, 204
Ladrão, the, 204, 206
Lagos, 123–4, 132
Lagosta, 120
Lamego, 126, 128
Lamiñak, 79, 81
Landim, 239
Lang, Andrew, 54
Larouco, 39, 77
Leão, Duarte Nunes de, 189, 251
Leça, Armando, 187, 262
Leça do Balio, 79
Leiria, 24, 130
León, 38
Lianor, Queen, 129
Ligares, 135
Lima, River, 40, 136
Linhares, Conde de, 102
Lisbon, 5–8, 16, 23, 25, 44, 55, 60, 62, 64, 73, 75, 86, 104, 106, 111, 112, 118, 119, 123, 127, 133, 137, 140–2, 161, 164, 165, 168, 175, 190, 199, 200, 215, 238, 245–65, 274, 280
Liz, River, 24
Lopes Dias, Jaime, 52
Lousa, 166
Lousã, 28, 159
Lugo, Council of, 214
Lundum, the, 252–4

Macedo de Cavaleiros, 138, 169
Macieira, 136
Madeira, 16
Madragoa, 7, 140–2, 263
Mafra, 11, 17, 161
Maia, 123, 180
Malhão, Serra de, 14
Malhão, the, 193, 204–5
Mangualde, 30
Mannhardt, 113, 151, 152
Manoel I of Portugal, 252, 277
Manoeline Art, 12, 17, 21–2
Manteigas, 32
Marão, Serra do, 34, 35, 36, 132
Marco de Canavezes, 72, 93
Maria I of Portugal, 253
Marranos, 36
Marriage, 88–94
Martins, Padre Firmino, 63
May Day, 116, 121–5, 181, 182
Maypoles, 122–3, 145
Meimôa, 211
Menano, Antonio, 261
Merceana, 128
Mercês, 89, 137, 138
Mickle, 47

Minho, River, 23, 163
Minho province, 27, 33, 40–5, 66, 77, 78, 85, 92, 93, 97, 102, 116, 118, 120, 123, 132, 133, 137, 144, 150, 183, 208, 231, 250
Minoan art, 11, 158
Miranda do Douro, 37, 117, 169–71, 183, 238, 278
Miranda, terras de, 38, 43, 277
Mirandela, 36, 238
Modinha, the, 252–4
Mogadouro, 37, 169
Moimenta da Beira, 30
Monção, 163
Monchique, Serra de, 15, 102, 115
Moncorvo, 36, 135
Mondego, River, 26, 28, 29, 243, 261
Monfurado, Serra de, 13, 60, 61, 78, 92, 112
Montalegre, 39, 122, 163
Montejunto, 17
Montijo, 118, 120
Montemor o Novo, 9, 10, 77, 113, 278
Montserrat, 17
Moon, lore relating to the, 60, 61, 65, 68, 102, 240, 279
Moors, influence of, and lore relating to the, 12, 13, 14, 15, 16, 17, 42, 72, 77, 81, 86, 163, 166–8, 175, 176–9, 196–8, 203, 249, 250, 251, 266
Morand, Paul, 249
Morris Dance, the, 167–8, 171
Mouraria, 7, 245, 263
Mouras Encantadas, 77–80, 151, 167, 183
Mourisca, 164–6, 167, 176
Mouriscada, 171–5, 179
Muge, 20–1
Mumming plays, 109–11, 175
Murça das Panoias, 35
Murray, Miss M. A., 57
Musset, Alfred de, 231

Navarre, 102, 130
Nazaré, 24, 134, 160, 161
Negrões, 278
Nelas, 73
Nereids, 79
New Year, 101–3, 184, 185
Nine, use of the number, 66, 68, 85
Niza, 12, 118
Nogueira, Serra de, 36
Nogueirinha, 169
Noronha, Eduardo, 253
Nunez, Airas, 194
Nuts-in-May, 125

Obidos, 23, 131, 182
Ochagavia, 130
Oleiros, 125
Olhão, 14
Oliveira, 72
Oliveira de Azemeis, 184
Oporto, 28, 29, 44–5, 53, 71, 72, 73, 75, 98, 125, 164, 166, 172, 175, 180, 228, 263
Outra Banda, 7–9, 16
Ovar, 7, 23, 29, 206
Oxen, lore relating to, 117, 128–9, 132, 150

Pais do Vinho, 34
Palha, 195
Palm Sunday, 148
Palmela, 8, 9
Parada de Infanções, 93
Paredes da Beira, 146, 152
Paredes de Coura, 144
Pedras Salgadas, 163
Pedro I of Portugal, 24, 28
Pedro II of Portugal, 106
Pedrogão Pequeno, 167
Pedroso, Consiglieri, 266
Pena Palace, 17
Penafiel, 117, 163
Penamacôr, 31, 162, 208, 211, 213
Penedones, 94
Penha d'Aguia, 87
Penha Longa, 127
Peniche, 23
Peninha, 17, 136
Pesqueira, 34
Philip II of Spain, 17
Philip III of Spain, 175
Philippa of Lancaster, 16
Phoenicians, traces of the, 7, 9, 23–4
Pillories, 35–6
Pimentel, A., 247, 253, 254, 265
Pinhel, 31, 32
Pinho Leal, 157, 163, 167
Pires, Tomás, 60, 231
Plays, traditional, 176–80
Pombal, 25, 89
Ponte da Barca, 40, 85
Ponte de Lima, 42, 60, 103, 136, 162, 179
Popular Saints (see also St Anthony, St John, St Peter), 123, 138–54
Portimão, 15, 123
Port wine, 34
Pottery, peasant, 11, 157
Póvoa, Romaria of, 137, 138, 156, 201, 209
Póvoa de Varzim, 23, 134
Pragança, 118

Praia da Rocha, 15
Provence, 190, 194–5, 228
Proverbs, 37, 38, 62, 77, 273–80

Quatrains, popular, 89, 90, 94, 119, 133, 135, 146, 153–6, 157, 159, 184, 193, 230–44
Quintanilha, 273

Red, lore relating to the colour, 85, 238
Refojos, 162
Regoa, 34, 75
Ribatejo, 18–21, 128, 206, 273
Ribeiro, 1, 40
Ribera y Tarragó, Julián de, 196–7, 203
Rocio Station, 5, 22
Roderic the Goth, 31
Rogations, 51, 162
Romans, influence of, and lore relating to the, 9, 23, 30, 39, 41, 77
Romarias, 38, 126–7, 155–81, 277
Roth, Cecil, 37

Sabugal, 31
Sagres, Cape, 15
St Anthony (see also Popular Saints), 24, 123, 132, 133, 134, 135
St Cyprian, book of, 72
St Helena, 75
St Hilary, 98
St John the Baptist (see also Popular Saints), 131, 133, 162, 220
—— festival of, 52, 78, 100, 122–3, 164–7, 171–5, 182, 184, 210, 213, 225, 250, 259, 279
St Mark, 87, 117
St Martin of Braga, 214
St Peter (see also Popular Saints), 11, 18, 271, 272
St Stephen, festival of, 103–4
St Thomas, 70
St Vincent, Cape, 14, 15
Salazar, Adolfo, 248
Saliva, lore relating to, 62, 68
Sampaio, Professor, 250, 259
Santa Isabel do Monte, 97
Santa Leocadia de Baião, 55
Santa Victoria do Ameixial, 92
Santarem, 18, 20, 21, 278
Santiago de Compostela, 97, 192, 195, 196, 213
Santo Tirso, 51, 111, 125
São Antão, 130, 133
São Bartolomeu do Mar, 88, 135
São Gonçalo de Amarante, 87, 131, 133, 166

São João das Lamparas, 145
São João do Campo, 43
São Martinho de Bougado, 53, 61
São Miguel, Serra de, 14
São Pedro de Rates, 131-2
São Pedro de Sarrazenos, 169
São Tiago da Cruz, 93
São Tiago de Cacem, 158
Sardão, 71
Sarmento, Martins, 80
Saudades, 190, 262-3
Sawing the old woman, 119-21
Segura, 135
Serpa, 206, 208, 210, 232
Serpents, lore relating to, 78-80, 151
Sertorius, 11
Setúbal, 8-9, 21, 164, 200
Seven, use of the number, 68, 144, 219, 223
Severa, 199, 263-4
Sezimbra, 8
Silius Africanus, 198
Silves, 14, 15, 198
Sines, 91, 123
Sitwell, Sacheverell, 126
Smith, Professor Elliot, 168
Soajo, Serra de, 43, 99
Sobrado, 171-5
Sotavento, 14
Soule, 176
Souto das Neves, 176
Spinning, lore relating to, 56, 77, 78, 80, 93, 118, 122
Stars, lore relating to the, 68
Stick dance, 169-71
Stolen articles, lore relating to, 136
Stones, lore relating to, 127, 128, 155
Stuart, Lord, 190
Sun, lore relating to the, 68, 69, 100, 102, 240

Taboa, 57, 132
Tafall, Santiago, 194
Tagus, River, 5, 7, 18, 31, 33, 161
Taveiroos, Pay Soarez de, 190
Tavora, 14
Teixoso, 52
Terroso, 159
Thirteen, use of the number, 68
Three, use of the number, 52, 68, 88, 95, 147, 216, 218
Thunderbolts, 59, 60, 65, 149
Tiles, lore relating to, 85, 136, 163, 238
Tinalhas, 125
Tolentino de Almeida, Nicolau, 253
Tomar, 21, 22, 63, 130
Tomás, Pedro Fernandes, 197, 259

Tondela, 73, 273
Torreira, 135
Torres Vedras, 17, 23, 109, 127, 128, 169, 226
Tourem, 120
Trancoso, 31, 93, 125
Travanca, 149
Trás-os-Montes, 11, 33, 34-40, 63, 71, 87, 92, 115, 120, 122, 124, 138, 162, 163, 171, 193, 202, 208
Travessó, 163
Trend, Professor J. B., 207, 214
Troubadours, 156, 189-96, 217, 228
Tua, River, 238
Turquel, 78
Twelfth Night, 103-4, 169, 180

Urgeiriça, 29, 32
Urtell, Hermann, 83

Valença do Minho, 111
Valerius Maximus, 44
Valladolid, Council of, 197
Valongo, 171
Vasconcelos, Antonio de, 189
Vasconcelos, Carolina Michaelis de, 196
Vasconcelos, José Leite de, 56, 58, 59, 74, 76, 82, 98, 124, 145, 150, 180, 193, 267
Vendas Novas, 9
Vermoil, 125
Vernoim, 78
Verstegan, Richard, 81
Viana do Castelo, 35, 40, 41, 42, 116, 176, 206, 232
Vicente, Gil, 22, 32, 59, 120, 159, 192, 242, 251, 274
Vieira, Ernesto, 252, 259
Vigo, 191, 194
Vila Cova, 82, 163
Vila do Bispo, 15, 152
Vila do Conde (Minho), 42
Vila do Conde (near Chaves), 180
Vila Franca da Xira, 20, 206
Vila Franca do Rosario, 111
Vila Nova da Gaia, 34, 44, 72
Vila Real, 11, 35, 135
Vila Real de São Antonio, 14
Vila Velha de Rodão, 242
Vilar Formoso, 4
Vimioso, 37
Vimioso, Conde de, 264
Vindel, Pedro, 192, 194
Vinhais, 60, 96, 115, 119, 121, 180
Vinho verde, 41, 160
Viriato, 30

Viseu, 30, 79, 272
Vivianez, Pero, 156
Vouga, River, 29
Vouzela, 125

Wassailing, 101–3, 182, 184
Water, lore relating to, 68, 69, 73,
    75, 79, 83, 85, 86, 96, 100, 102,
    135, 144, 145, 146, 151–2, 162,
    174
Wellington, Duke of, 29, 37, 44
Werewolves, 71, 81–3, 151, 183
Westermarck, 152
Whitsuntide, 104, 134, 157, 158

Witchcraft, 55–7, 61, 65, 66, 68, 71,
    76, 81–3, 86–7, 125, 151–2, 173,
    183, 185
Wolfram, Dr Richard, 168, 184
Wortley Montagu, Lady Mary, 7

Yokes, decorative, 44
Young, George (now Sir George), 1

Zefa, 63
Zejel, the, 196
Zezere, River, 21
Zorro, João, 194
Zuloaga, Ignacio, 31